Intelligent Spectrum Management

IEEE Press
445 Hoes Lane
Piscataway, NJ 08854

Intelligent Spectrum Management

Towards 6G

Edited by

Sridhar Iyer
KLE Technological University, Dr MSSCET, Belagavi, Karnataka, India

Anshuman Kalla
Department of Computer Engineering, CGPIT, Uka Tarsadia University (UTU), India

Onel Alcaraz López
Centre for Wireless Communications (CWC), Oulu, Finland

Chamitha De Alwis
School of Computer Science and Technology, University of Bedfordshire, United Kingdom

IEEEPress
WILEY

Published by John Wiley & Sons, Inc., Hoboken, New Jersey.
Published simultaneously in Canada.

Library of Congress Cataloging-in-Publication Data applied for:

Hardback ISBN: 9781394201204

Cover Design: Wiley
Cover Image: © tanit boonruen/Getty Images

Set in 9.5/12.5pt STIXTwoText by Straive, Chennai, India

Contents

About the Editors

Sridhar Iyer (Senior Member IEEE) received the MS degree in electrical engineering from NMSU, USA, in 2008, and PhD degree from Delhi University, India, in 2017. He received the young scientist award from the DST/SERB, Govt. of India, in 2013, and Young Researcher Award from the Institute of Scholars in 2021. He is the recipient of the "Protsahan Award" from IEEE ComSoc, Bangalore Chapter, as a recognition to his contributions toward paper published/tutorial offered in recognized conferences/journals during 2021, 2022, and 2023. He has completed two funded research projects as the principal investigator and is currently involved in an ongoing funded research project as the principal investigator. He is currently an Editorial Board Member of Scientific Reports, Nature. His current research focus includes semantic communications and spectrum enhancement techniques for intelligent wireless networks. Currently, he serves as professor at KLE Technological University, Dr MSSCET, Belagavi, Karnataka, India. For more info visit https://scholar.google.co.in/citations?hl=en&user=2hbORHEAAAAJ&view_op=list_works&sortby=pubdate.

Anshuman Kalla (Senior Member IEEE) is working as professor at the Department of Computer Engineering, CGPIT, Uka Tarsadia University (UTU), India. Dr. Kalla has more than 15 years of teaching and research experience. He has worked as a postdoctoral visiting researcher at the Center for Wireless Communications (CWC), University of Oulu, Finland. He graduated as an engineer from Govt. Engineering College Bikaner in 2004. He did Master of Science in telecommunications and wireless networking from ISEP, Paris, France, in 2008 and another master's from University of Nice Sophia Antipolis, France, in 2011. He obtained a PhD degree in 2017. Dr. Kalla was recipient of master's scholarships for pursuing both his master's programs. Dr. Kalla has delivered invited sessions and talks during various short-term courses, faculty development programs, workshops, and conferences. In particular, he has delivered tutorials at IEEE 5G World Forum in 2021, IEEE ANTS 2020, and IEEE 5G World Forum in 2020. He has published papers in reputed international journals such as Elsevier (*JII*, *IPM*, *JNCA*, *COMNET*, and *ICT Express*), *IEEE Consumer Electronics Magazine*, *IEEE ComSoc*, *IEEE Computer*, *IEEE Potentials*, and *IEEE EMR*. His areas of interest are blockchain, 5G, 6G, IoT, information centric networking, software-defined networking, and next-generation networks. For more info visit https://sites.google.com/site/kallanshuman.

Onel Alcaraz López (Senior Member IEEE) is an associate professor (tenure track) in sustainable wireless communications engineering at the Centre for Wireless Communications (CWC), Oulu, Finland. He received the BSc (1st class honors, 2013), MSc (2017), and DSc (with distinction, 2020) degrees in electrical engineering from the Central University of Las Villas (Cuba), the Federal University of Paraná (Brazil), and the University of Oulu (Finland), respectively. From 2013 to 2015, he served as a specialist in telematics at the Cuban telecommunications company (ETECSA). He is a collaborator to the 2016 Research Award given by the Cuban Academy of Sciences, a co-recipient of the 2019 and 2023 IEEE European Conference on Networks and Communications (EuCNC) Best Student Paper Award, the recipient of both the 2020 best doctoral thesis award granted by Academic Engineers and Architects in Finland TEK and Tekniska Föreningen i Finland TFiF in 2021, and the 2022 Young Researcher Award in the field of technology in Finland. He is co-author of the books entitled *Wireless RF Energy Transfer in the Massive IoT Era: Towards Sustainable Zero-Energy Networks*, Wiley, December 2021, and *Ultra-Reliable Low-Latency Communications: Foundations, Enablers, System Design, and Evolution Towards 6G*. He is currently an associate editor of the *IEEE Transactions on Communications*. For more info visit https://sites.google.com/view/onellopez.

Chamitha De Alwis (Senior Member IEEE) is a lecturer at the School of Computer Science and Technology, University of Bedfordshire, United Kingdom. He received the BSc degree (First Class Hons.) in electronic and telecommunication engineering from the University of Moratuwa, Sri Lanka, in 2009, and the PhD degree in electronic engineering from the University of Surrey, United Kingdom, in 2014. He has secured several competitive research grants, actively contributed to many research projects, delivered guest talks and tutorials, and provided professional consultancy services to the industry. He actively pursues research in areas including AI security, cybersecurity, network security, 5G/6G, and blockchain.

For more info visit https://sites.google.com/view/chamitha.

Foreword

The vision for a fully digitized, data-driven, and autonomous society in the next decade demands future communication networks to provide ubiquitous, high-performance connectivity and thus facilitate advanced applications and services. Recently, the basic 5G standardization was completed, marking a significant milestone and paving the way for potential enhancements Beyond 5G (B5G) networks and technologies, which are currently under study. Indeed, the research community has already envisioned that by 2030, a reinvented mobile communication will be required to support the next generation of bandwidth-hungry, ubiquitous, and intelligent services and applications. Therefore, both the industry and the research community have already begun conceptualizing 6G to sustain the competitive edge of wireless technology and meet the communication requirements of the next decade. Further, effectively utilizing the broad spectrum bands in the next-generation mobile networks, including mitigating existing and future spectrum crunch issues, will require devising new intelligent and dynamic spectrum management techniques. Various techniques such as cognitive radio, blockchain, deep learning, and reinforcement learning can be used to enable intelligence within the wireless networks. In turn, this will result in an increased utilization of the spectrum, allowing networks to meet the heterogeneous demands of the users, thereby fostering the technological and market growth of the wireless communication industry.

This book aims to delve into efficient spectrum management for the next-generation intelligent mobile networks. For this, key technologies, the convergence of AI with advanced techniques, and physical and virtual infrastructure management are comprehensively discussed. Initially, the book introduces the reader to the topic through detailed information on the evolution of mobile networks, including the evolution of spectrum management toward 6G. Then, specific technologies for spectrum management, such as deep learning, blockchain, and reinforcement learning, are thoroughly discussed. Further, aspects related to the convergence of artificial intelligence (AI) and blockchain for spectrum management are presented, followed by a discussion on multiple techniques, such as network slicing and virtualization for implementing dynamic and decentralized spectrum management in the 5G/B5G/6G networks.

The book is arranged in the following four sections:

1. **Section I** introduces the readers to the various generations of wireless networks followed by the aspects related to spectrum management and presents various options to access the spectrum in 6G networks. The section includes contributions titled "Evolution of Mobile Networks" and "Spectrum Access Options for Local 6G Networks."
2. **Section II** details the various techniques that can be implemented for spectrum management in the wireless networks. The section includes contributions titled "Spectrum Management Technologies in Mobile Networks," "Artificial Intelligence Enabled Dynamic Spectrum Management," and "Infrastructure for Spectrum Management Enabled by Virtualization and Network Slicing."
3. **Section III** details the implementation of reflective intelligent surfaces (RIS) for spectrum management in 6G networks. The content presented also includes specific case studies for validating the implementations. The section includes contributions titled "Spectrum Management for 6G RIS-SWIPT Systems," "Reinforcement Learning and Deep Learning Assisted Spectrum Management for RIS-SWIPT Enabled 6G Systems," and "RIS-Aided Low Complexity Waveform Design for Joint Sensing and Communications."
4. **Section IV** includes the implementation of blockchain for spectrum management followed by the joint implementation of blockchain with AI for spectrum management. The section includes contributions titled "Blockchain and Smart Contracts for Decentralized and Secure Spectrum Management Toward 6G—Beyond Hype" and "The Synergy of Artificial Intelligence and Blockchain in 6G Spectrum Management."

Through the aforementioned sections of the book, the readers will find comprehensive discussion on:

1. identification of the key technologies and techniques to enable efficient spectrum management in the 5G/B5G/6G intelligent wireless networks.
2. aspects related to physical and virtual infrastructure for spectrum management.
3. mutual benefits, challenges, and prospective research on AI and blockchain for spectrum management.
4. blockchain and smart contracts for decentralized and secure spectrum management toward 6G.

The book may be of key interest to:

1. **Researchers**: This book sheds light on the technologies for spectrum management in future wireless networks. To those initiating research in this area,

the book can help in identifying the innovative methods that can be adopted to ensure increased spectrum utilization and management. The book also provides a comprehensive treatment of the principles and architectures for spectrum management. Overall, for the researchers, this book can provide them with a clear understanding of spectrum management in intelligent mobile networks and can help them develop future related frameworks.

2. **Academicians**: The book may be valuable for the academicians who engage in teaching and conducting research in the area related to intelligent wireless networks.
3. **Network Service Providers**: This book can help the service providers gather key information regarding spectrum management. This can help them plan future frameworks and algorithms.
4. **Industry Professionals**: Industry professionals may use this book as a reference to envision, plan, and implement future applications and use cases considering appropriate business models.
5. **Regulators and Standardization Bodies**: For regulators and Standards institutions, this book can help to inform them about forthcoming technologies and applications.

Preface

Fifth-generation (5G) mobile networks are expanding, underpinning applications ranging from smart cities and smart industries to autonomous vehicles. However, as researchers are exploring emerging applications, including augmented reality (AR), virtual reality (VR), mixed reality (XR), holographic telepresence, connected autonomous vehicles (CAVs), Industry 5.0, and Metaverse technologies, it is evident that the world demands better connectivity that can be realized through Beyond 5G (B5G) technologies, developing toward the sixth generation (6G) of mobile networks. The 6G networks will exploit a plethora of new technologies and harness the capabilities of artificial intelligence (AI) to power the next-generation applications.

The B5G and 6G networks are expected to manage heterogeneous communication technologies utilizing a broad range of frequency spectrum up to terahertz (THz) frequencies. This demands efficient and intelligent spectrum management. In this direction, the book provides a comprehensive understanding of the advances in the research and development work related to B5G and 6G spectrum management in wireless networks. The book explores the evolution of mobile networks focusing on spectrum management technologies in mobile networks (principles, technical aspects, applications, and challenges). Furthermore, the book provides a detailed understanding of the evolution of spectrum management toward 6G, together with various technologies such as deep learning, reinforcement learning, and blockchain, which enable efficient spectrum management. The book also elaborates on the convergence of AI and blockchain to facilitate intelligent and secure spectrum management in the B5G and 6G networks. In addition, the book highlights the infrastructure requirements for realizing dynamic spectrum management in the B5G and 6G wireless networks. Therefore, it is envisaged that the book will be a valuable resource for the communication researchers, academics, industry experts, and standardization bodies working toward realizing intelligent spectrum management solutions for the B5G and 6G networks.

15 October 2024

Sridhar Iyer
KA, India
Anshuman Kalla
GJ, India
Onel Alcaraz López
Oulu, Finland
Chamitha De Alwis
Luton, UK

Acknowledgments

In compiling this book, we have been fortunate to receive state-of-the-art technical contributions from renowned researchers. This book could not have been completed without their contributions. We are deeply indebted to the following researchers who took the time to frame and revise their contributions:

Deepak Kumar, Marja Matinmikko-Blue, Seppo Yrjola, Petri Ahokangas, Harri Saarnisaari, Qiyang Zhao, Hang Zou, Yu Tian, Lina Bariah, Belkacem Mouhouche, Faouzi Bader, Ebtesam Almazrouei, Merouane Debbah, Uditha Wijewardhana, Nishan Dharmaweera, Bhathiya Pilanawithana, Neha Sharma, Sumit Gautam, Prabhat Kumar Upadhyay, Symeon Chatzinotas, Bjorn Ottersten, Manojkumar B. Kokare, Purva Sharma, Swaminathan R., Vimal Bhatia, Christos Tsinos, Soumya P. Dash, Aryan Kaushik, Aakash Arora, Marco Di Renzo, Bikramjit Choudhury, Pranav Kumar Singh, Panchanan Nath, Ujjal Roy, Ramalingam M., Gokul Yenduri, Pyingkodi M., and Thippa Reddy Gadekallu.

We would like to thank Kavipriya, Becky, and Sandra from Wiley for their constant support and timely inputs regarding the completion of various phases of this book. We also thank Wiley for their assistance in creating the index for the book. We are thankful to our respective institutions for providing us the opportunity to carry out this work and for providing us with an intellectually stimulating environment.

As always, the greatest debt one owes is to one's colleagues, friends, and family. In our case, this debt is especially large. So, we would like to acknowledge the unwavering support by our family members who are responsible for this *book* in more ways than even they know. We dedicate this book to them.

Sridhar Iyer
Anshuman Kalla
Onel Alcaraz López
Chamitha De Alwis

Section I

1

Evolution of Mobile Networks

Deepak Kumar[1], Sridhar Iyer[2], and Onel Alcaraz López[1]

[1] *Centre for Wireless Communications (CWC), University of Oulu, Oulu, Finland*
[2] *Department of ECE, KLE Technological University, Dr. MSSCET, Belagavi, Karnataka, India*

1.1 Introduction

The advent of mobile networks has ushered in a new era of communication. This in turn is playing a pivotal role in our interconnected world and has transformed the way we connect, work, and live. By enabling people to seamlessly communicate, access information, and utilize digital services anytime and anywhere, mobile networks increasingly bridge the digital divide and support businesses, e-commerce, and digital entrepreneurship, thus fueling economic growth and societal advances. Indeed, mobile communications can significantly contribute to achieving the sustainable development goals of the United Nations.[1] This is possible by offering infrastructure and access to digital services supporting growth, efficiency, and sustainability, especially for economies where existing services are limited or the related infrastructure is poor [1].

Digitalization of services delivered through mobile communications networks has shown immense benefit in developing economies in particular driving the uptake of micro-banking and micro-finance, micro-energy grids, and market creation [2]. Productivity enhancement may arise from remote work support and access to critical information and digital services, a capability that has become crucial in the face of global challenges such as the COVID-19 pandemic. In times of crisis, mobile networks even serve as vital lifelines, facilitating emergency communication and coordination.

There have been phenomenal advancements in mobile communications for more than a century and consistently since the first-generation (1G) networks

1 Refer to https://sdgs.un.org/goals.

Intelligent Spectrum Management: Towards 6G, First Edition.
Edited by Sridhar Iyer, Anshuman Kalla, Onel Alcaraz López, and Chamitha De Alwis.

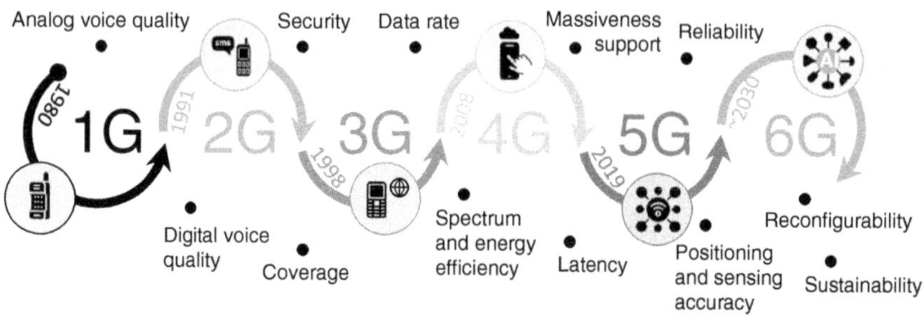

Figure 1.1 Evolution of mobile networks and emerging KPIs.

in the 1980s. A new generation has been introduced nearly every 10 years since then, with each iteration bringing significant improvements in terms of key performance indicators (KPIs). From the initial 1G networks that allowed only voice calls, mobile communication networks have progressed through the second generation (2G), third generation (3G), fourth generation (4G), and fifth generation (5G), enabling not only voice communication but also high-speed Internet access, mobile gaming, location tracking, online education, augmented reality applications, and foundation for the Internet of Things (IoT) [3]. A graphical representation of the evolution of mobile networks, including some distinctive KPIs that emerged during the evolution, is shown in Figure 1.1.

Understanding the evolution of mobile networks is of paramount importance for researchers, engineers, and academics, and constitutes the scope of this chapter. Such an understanding/knowledge drives innovation, informs policy, opens business avenues, and ensures the education of future generations. As mobile networks continue to evolve, staying informed and engaged in their development is vital for the betterment of society and the advancement of human, human-to-machine, and machine-to-machine connectivity.

1.2 Origins and Early Developments

Telecommunications have been critical to human society since ancient times. The discovery of the relationship between electricity and magnetism by Hans Christian Oersted in 1820 was a milestone. Further, Michael Faraday showed that a fluctuating magnetic field could induce electric current on a conductor in 1831, marking a turning point in the history of telecommunications by laying the foundation for subsequent advancements. This discovery laid the groundwork for wireless communication at non-line-of-sight distances.

Later on, two major telecommunication developments were the invention of the telephone by Alexander Graham Bell in 1876, allowing the transmission of analog live signals, and Marconi's wireless telegraphy in 1896. The first amplitude-modulation radio broadcasting was demonstrated by Reginald A. Fessenden in 1906. Meanwhile, in 1935, Major Edwin H. Armstrong demonstrated high-quality music transmission using frequency-modulation radio broadcasts. The private telephone companies also started providing landline telephones and services. During the 1940s, the general public was offered a public mobile telephone, which was a wireless device that connected with the public switched telephone network [4]. The United States' Federal Communication Commission (FCC) started two-way radio service over a 460 MHz band in 1945, where millions of users used the same channel across the country. In 1947, radiotelephone systems, comprising small geographical areas called cells, were proposed by the Bell laboratories [5]. A base station (BS) transmitter setup was placed in each cell, and cell traffic was controlled by a central switch. A few decades later, in 1971, the first wireless computer network, called AlohaNet, was introduced [6]. AlohaNet could connect multiple low-data rate stations via a single radio channel to a central host without considering any access rule or synchronization [7]. In 1972, S-Aloha was proposed using a time-slotted channel, which could double the channel capacity. The cellular radio networks assign radio channels to mobile stations by employing a demand-assigned multiple-access protocol, where the uplink request channel is based on S-Aloha.

The initial cellular systems are referred to as mobile radiotelephone or zeroth generation (0G), pointing to the pre-cellphone mobile telephony. Such systems were usually mounted in vehicles' boot/trunk. Indeed, the transceiver was mounted in the vehicle boot and was usually placed on the head section and fixed close to the driver's seat. In 1979, Nippon Telegraph and Telephone Corporation launched 1G to the citizens of Tokyo, which was available nationwide in Japan by 1984 [8]. The cellular standards used in 1G included mainly Advanced Mobile Phone Services (AMPS) and Nordic Mobile Telephones (NMT). The AMPS was invented at Bell Labs [9] and also used in the United Kingdom under the name of Total Access Communication System [10]. The NMT system was simultaneously introduced in Denmark, Finland, Norway, and Sweden in 1981, which was the first mobile phone network with worldwide roaming [11]. Motorola's DynaTAC mobile phone was implemented by Chicago-based Ameritech in 1983 to build the first 1G network in the United States [12]. The world's first cell phone, DynaTAC 8000X, created by Motorola in 1983, was essentially a two-way frequency-modulation radio designed only for voice calls [13]. Several nations, including the United Kingdom, Mexico, and Canada, followed in the early to mid-1980s.

1G was completely an analog cellular system. The fundamental concept underlying 1G cellular networks is the division of the region into cells, each of which is serviced by a BS and usually spans 10–25 km. The small size of cells allows for frequency reuse in nearby cells. In addition, smaller cells need smaller, less-expensive, and less-powerful equipment to send and receive data. However, these standards did not include any placement (i.e., coordinates of a certain object) instructions. The utilization of location data within the network garnered noteworthy interest from both, operators and application developers. Vehicle location (by using signal strength, time delay, or direction of arrival measurements) was targeted to improve the effectiveness of cellular calls through system control [14]. 1G cellular systems were additionally utilized for intelligent vehicle highway system (IVHS) applications [15]. Further, it supported emergency services based on proprietary location solutions. This included services such as those of Grayson Wireless with a joint time difference of arrival and angle of arrival solution or true position with an uplink time difference of arrival, both of which used AMPS signals [16]. 1G networks realized data rates up to 2.4 kbps, sufficient to support analog voice call services [10]. Meanwhile, these networks were concentrated in urban and densely populated areas with limited coverage and capacity, and faced network congestion, excessive call drop problems, reckless hand-off, and security threats.

The Groupe Spécial Mobile, established in 1982 by the European Conference of Postal and Telecommunications Administrations, started working on the harmonization of mobile communication systems in the 900 MHz band [9]. The primary service was telephony; however, it was also possible to use enticing nonvoice alternatives. The action plan included studies on market, tariff, technical, and regulatory needs. Further, in 1984, France and Germany contributed to the global system for mobile communications (GSM) [17]. A structure for the GSM specifications and an action plan for their completion was created. It was decided to employ smart cards as subscriber identity modules (SIMs), and support free circulation, mobile station licensing, international roaming, charging, and accounting. At a high-level meeting in Bonn in May 1987, the basic GSM specification was finally agreed upon.

The transition from 1G to 2G in the 1990s constituted the transition from analog to digital cellular systems, offering improved voice quality and paving the way for sophisticated mobile services and capabilities. The progression of mobile communications from several separate systems to standard systems across the nations was one of the primary factors that made this transition possible. The 2G system aimed to enhance spectrum efficiency while offering a fixed data rate of 64 kbps [18]. Additionally, 2G systems offered features like electronic mail (e-mail) and short message service (SMS). The first SMS message "Merry Christmas," was sent by Neil Papworth, a test engineer for Sema Group, from a computer to his colleague Richard Jarvis's Orbitel 901 phone on December 3, 1992 [19]. In 1993, Nokia's mobile phones were the first to support consumer SMS texting.

The major 2G cellular system standards were GSM, interim standard 95 (IS-95), and interim standard 136 (IS-136). The physical layer of GSM is based on time-division multiple access (TDMA) and frequency-division multiple access (FDMA). The TDMA of GSM was followed by digital standards such as digital AMPS or IS-54 (later on substituted by IS-136), integrated dispatch enhanced network (iDEN), and personal digital cellular [8]. The code-division multiple access (CDMA) was added to the IS-95 standard to improve the capacity of mobile cellular networks. CDMA also provided support for multiple data rate sets. The GSM's standard was split into two phases, enabling first, rapid implementation of common services such as telephone or SMS, and then, the introduction of technical advancements and additional services [20].

Notably, there were no positioning/location mechanisms within the standard. The positioning capability of GSM was limited to the use of training or synchronization signals to compute ranging requirements, and similar procedures were employed for the CDMA system. In 1996, the FCC approved the provision of enhanced 911 (also known as E-911 or E911, which was a system that gives 911 dispatchers the caller's location automatically in North America) for location requirements on the existing TDMA and CDMA cellular systems. Later on, global positioning system (GPS) receivers were incorporated to assist the cellular networks. Meanwhile, digital cellular networks were evolving toward GSM phase 2+ (i.e., 2.5G and 2.75G) and universal mobile telecommunications systems (UMTS). The 2.5G comprises a general packet radio system (GPRS) with a data rate of 50 kbps, and 2.75G comprises an enhanced data rate for GSM evolution (EDGE) with a data rate of 200 kbps, including packet-switched services like transport control protocol (TCP)/internet protocol (IP). The third generation partnership project (3GPP) was created in 1998 as a partnership of international members to standardize the evaluation of GSM and UMTS, with the European Telecommunications Standards Institute (ETSI) being one of the contributors and the main sponsor. In parallel, the Telecommunications Industry Association (TIA) and the Electronic Industries Alliance (EIA) established the 3GPP2 collaboration to carry out the standardization of IS-95 and CDMA2000 technologies.

Cellular networks were originally designed for circuit-switched speech communication and later offered data as an add-on. Data transfer via analog cellular telephone connections utilizing modems was a basic method of mobile data communication. For instance, analog and digital cellular networks such as EIA-553, AMPS, and ETSI GSM, provided modem-based circuit-switched transparent data service [7]. In this communication method, the network merely provides a voice link over which the mobile data modem can interact with a corresponding data modem in an office or computer center. The circuit-mode connection was largely underutilized, and the service was expensive if only brief messages were sent back and forth over the network during a (long) interactive session. This was the driving force for the creation of mobile data networks that use end-to-end packet switching. The packet switch was an evolution of the GPRS with a more optimized functional split between the UMTS terrestrial radio access (UTRA) network and the core network.

1G and 2G network evolution was driven by the need to support analog and then digital voice communications. Spectrum allocation was relatively static, with fixed frequency bands designated for specific services. 1G networks, e.g., AMPS and NMT, operated primarily in the 800–900 MHz range, utilizing analog signals and simple frequency division. This period saw significant regulatory involvement to allocate and license the spectrum to the operators, ensuring minimal interference and promoting initial market growth. With the advent of 2G networks, like GSM, there was a shift to digital transmission, significantly improving the spectral efficiency by exploiting TDMA and FDMA technologies. Spectrum efficiency became crucial as 2G introduced SMS, basic data services, and even digital encryption for improved security. This period laid the foundation for more dynamic and competitive spectrum management practices as mobile communication started to scale globally.

1.3 Data-Centric Mobile Networks

The previous generations of cellular communication systems were based on circuit and packet switching modes while supporting poor data rates, capacity, and limited services. The explosive growth of the Internet (entailing worldwide access to information) and multimedia applications would become the driving force for the evolution into data-centric mobile networks. A mobile Internet, i.e., using a mobile device to access the Internet over a wireless connection, represented realizing ubiquitous access to data communications. With this technology, users could finally access various online services and content without a traditional wired connection to the Internet.

The International Telecommunication Union established the International Mobile Telecommunications 2000 (IMT-2000) framework as an international standard for 3G cellular networks in 1998 [21]. 3GPP aimed to develop globally applicable specifications for 3G mobile systems. 3GPP standards were designed for mobile systems based on evolved GSM core networks, as well as the supported radio access technologies, service capabilities, and security. 3GPP was the standards organization behind UMTS/wideband CDMA (W-CDMA), and 3GPP2 developed separate standards for 3G technologies, known as CDMA2000 [22]. The emerging 3G systems enabled universal access to a variety of services tailored to mobile users as well as a wide range of basic and supplementary services supported by networks. The 3G architecture was evolved from 2G GSM, GPRS, and UMTS. Specifically, a set of extensions to GSM data services were designed and standardized, resulting in 2.5G systems viz., high speed circuit switched data (HSCSD), GPRS, and EDGE. The main 3G technologies were UMTS and CDMA2000, and are discussed below.

- The UMTS was a wideband CDMA (WCDMA) technology, also called UTRA. The frequency-division duplex (FDD) and time-division duplex (TDD) modes of the UTRA network were set up to produce data speeds of up to 2 Mbps. UMTS R99 allowed mobile networks to directly profit from the vast impetus behind the Internet's expansion and launch of new services. This evolution also enabled telecom providers to roll out a standard backbone (like IP) for all access types, significantly lowering capital and operating costs. Meanwhile, UMTS R00 introduced a next-generation network architecture for circuit and packet-switched domains. Both, stationary and mobile customers, were able to access a variety of services using UMTS. The World Administrative Radio Conference in 1992 designated the 1885–2025 MHz and 2110–2200 MHz frequency bands as the target radio interfaces for UMTS, which was designed to be extremely flexible and efficient.
- The CDMA2000 technology was based on the evolution of IS-95, which consists of multiple narrowband CDMA carriers of 1.25 MHz. To safeguard significant financial investment in the base of networks that have already been deployed, the design of the air interface was backward compatible with the deployed IS-95 networks and built on 2G. The cdma2000 supported an advanced medium access control feature to efficiently support several concurrent data (up to 2 Mbps) and voice services.

The progression of GSM, UMTS, and CDMA2000 toward 4G cellular systems aimed to support Internet browsing, interactive gaming, mobile TV, and streaming video and audio [23]. This was led by long-term evolution (LTE), which is regarded as a 3.9G technology since 3GPP Rel-8 [24] and Rel-9 [25] of the LTE standard do not completely comply with the International Mobile Telecommunications (IMT)-Advanced standards for 4G technology. The LTE technology was designated by 3GPP to support mobile data demand and multimedia applications. Its evolved-UTRA air interface uses single-carrier (SC) FDMA for uplink and orthogonal frequency-division multiple access (OFDMA) for downlink to achieve a data rate of up to 50 Mbps and 100 Mbps, respectively [26]. The multicarrier feature allows LTE to operate in different system bandwidths (between 1.4 and 20 MHz) by adjusting the number of subcarriers and their allocation to multiple users.

LTE operates in the FDD and the TDD modes. A time-frequency block is the fundamental LTE radio resource that can be addressed for data transfer. The resource block has a subframe duration of only 1 ms in the time domain. The inclusion of a short subframe allows for the exploitation of channel variations by scheduling users based on their channel quality. Further, it is feasible to achieve broadly scaled transport block sizes and a wide variety of user-data rates by assigning a variable number of resource blocks to a particular user and choosing a modulation and coding scheme to suit the present channel conditions. Furthermore, an LTE system was designed to operate as a packet-based system with fewer network components, which increases system capacity and coverage, and offers enhanced quality of service (QoS) in terms of high data rates, low access latency, flexible bandwidth operation, and seamless integration with other current wireless communication systems.

3GPP began establishing advanced LTE, aka LTE-A, standards in Rel-10 [27] and termed it 4G technology to fulfill the peak data rate of up to 1 Gbps and low data rate of up to 100 Mbps [28]. The heterogeneous networks (comprising of macro and small cells) were defined to fulfill new features, including carrier aggregation (CA), coordinated multipoint, and improved multiple-input multiple-output (MIMO) transmissions. With SC-FDMA for uplink and OFDMA for downlink, LTE-A's physical layer was backward compatible with LTE. The maximum system bandwidth of the LTE-A technology was 100 MHz. As a result, the operators were able to combine fragmented spectrum from several bands into one bigger spectrum resource. The implementation of CA networks has advanced well, enabling operators to convert their investment in more LTE carriers into commercially viable high data rates. Moreover, CA is considered a vital enabler for harnessing unlicensed spectrum opportunities within the 5 GHz band, as specified in 3GPP Rel-13 [29].

The LTE-A system improved the previous LTE systems by supporting significantly more data usage, lower latency, and better spectral efficiency. In addition, it introduced a new type of data communication between entities called machine-type communication (MTC). With MTC, machines can share and exchange data without any human intervention. In the vision of smart city applications such as online education, smart utility, online surveillance, environment monitoring, e-health, and connected vehicles, a large number of autonomously operated low-cost devices (i.e., actuators and sensors) are required to be connected to physical objects. MTC is a term used to describe the communications that take place between these independently operating devices. Contention-based random access is used to support MTC in LTE.

The further enhancements of the LTE standard in 3GPP Rel-13 are known as LTE-Advanced Pro and termed 4.5G. A fundamental component of this evolution is indoor location, which plays a crucial role in improving mobile broadband and connectivity QoS while maintaining full LTE-A backward compatibility. Radiofrequency pattern matching (RFPM) was adopted to improve the existing positioning techniques dependent on radio access technologies such as observed time difference of arrival (TDoA), uplink TDoA, and enhanced cell-ID. Additionally, systems independent of radio access technologies, such as advanced global navigation satellite systems, terrestrial beacon systems, and WiFi/Bluetooth, were adopted to improve existing positioning techniques. The goal of LTE-A Pro's evolution was to offer a fresh 5G candidate technology.

The transition to 3G and 4G networks greatly reshaped spectrum management due to increasing data demands. 3G networks, e.g., UMTS and CDMA2000, leveraged wider frequency bands around 2 GHz and introduced technologies like WCDMA and CDMA, enabling higher data rates and more efficient spectrum use. Spectrum auctions became common, promoting a more competitive environment and better resource allocation. With the advent of 4G, the focus shifted to high-speed data and low latency. Spectrum management strategies in this generation included dynamic spectrum access and CA, which allowed operators to combine multiple frequency bands to enhance bandwidth and improve user experience. Additionally, the introduction of OFDMA enabled a more efficient use of the spectrum through time/frequency-domain multiplexing. Regulatory bodies played a critical role in standardizing these practices and ensuring that the spectrum was used efficiently and fairly.

1.4 5G Mobile Networks

Cellular mobile radio's communication and positioning functions are undergoing a revolution owing to the 5G technology. With the advent of 5G, connectivity provision has experienced significant improvements in wireless data rate, capacity, and coverage, together with a reduction of round-trip latency and energy usage. The 5G wireless standard envisions a network that is always available everywhere. Millions of 5G handsets and hundreds of operators have been involved in the commercial deployment of the 5G mobile networks globally. 5G-capable devices are not only limited to smartphones as they may also be found in a wide range of different forms, including head-mounted displays, laptops, drones, and fixed wireless access. 3GPP Rel-15 [30] in 2016 established the groundwork for 5G new radio (NR) by considering a variety of service needs, a broad spectrum (ranging from hundreds of megahertz to tens of gigahertz), and multiple deployment scenarios (indoor/outdoor, macro/small cells). Both, standalone and nonstandalone 5G mobile network architectures, such as those utilizing dual connection between LTE and NR, were defined to make it easier for operators to transition from 4G LTE to 5G NR.

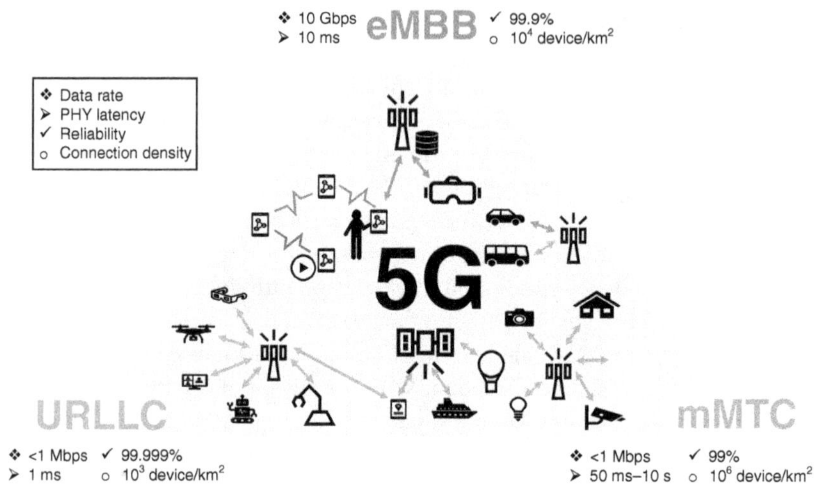

Figure 1.2 Capabilities of a 5G wireless network.

5G mobile networks are aimed to fulfill the recommendation of IMT for 2020, providing three main capabilities as illustrated in Figure 1.2 [31]:

- **Enhanced Mobile Broadband (eMBB)**: Hotspots and wide-area coverage for high throughput and a seamless user experience.
- **Ultrareliable and Low Latency Communications (URLLC)**: High availability and low latency.
- **Massive Machine Type Communications (mMTC)**: Very broad coverage and a large number of inexpensively linked devices with long battery life.

The 5G architecture offers better software services and network solutions for a variety of public and private sectors, including energy, agriculture, city administration, health care, manufacturing, and transportation. The device-centric feature of 5G is anticipated to make it possible for nearby devices to directly communicate while skipping cellular BS to share pertinent resources. Such MTC involves automated data generation, processing, transfer, and exchange amongst intelligent devices, with minimum human participation. This is a component of the IoT ecosystem, which envisions millions of simultaneous connections between various gadgets, connected homes, smart grids, and intelligent transportation systems. The 5G wireless technology's high bandwidth, widespread availability, and low latency are ensuring smart and intelligent vehicular communications. Technology developments in sensing and communication have created new opportunities for health monitoring. A paradigm shift in real-time remote patient health monitoring is made possible by the emerging 5G mobile networks and body area networks.

The 5G spectrum is divided into multiple frequency ranges (FRs). Specifically, FR1 corresponds to the region of 0.410–7.125 GHz, while FR2 covers 24.25–52.6 GHz (FR2-1) and 52.6–71 GHz (FR2-2). 5G mobile networks aim to provide a data rate of 10 Gbps in uplink and 20 Gbps in downlink, a 10-fold increase from the highest data rate of 150 Mbps for LTE cellular networks. Also, a latency below 1 ms is targeted, which is also a 10-fold improvement over the 10 ms round-trip duration of 4G mobile networks. To support an enormous number of linked devices as well as ultrahigh bandwidths for extended periods in a specific space, 1000 times greater bandwidth per unit area is required [32]. Moreover, longer battery life (up to 10 years) is supported to accommodate these and other emerging applications.

5G uses beam division multiple access (BDMA) and non- and quasi-orthogonal or filter bank multicarrier (FBMC) multiple access [33]. Within the BDMA framework, an orthogonal beam is assigned to each mobile station based on its position such that the mobile stations can receive numerous accesses, thus increasing the system's capacity [34]. Notably, the beamforming capabilities at the transmitter and receiver and the exploitation of spatial division multiple access allow greater frequency reuse.

Interestingly, the mobile industry's rising demand is pushing the transition from traditional cellular coverage to cell-free device-centric deployments. This can be realized by exploiting large beamforming gains, high frequency/bandwidth, and interference-taming capabilities of distributed massive MIMO (M-MIMO) networks. Note that millimeter-wave (mm-wave) propagation's small radio wavelengths make it possible to employ numerous small antennas. By utilizing array antennas to control the signal's phase and amplitude, electromagnetic waves can be enhanced in the desired direction while being canceled out in all other directions. Further, smart antennas support interference mitigation while preserving the best coverage area and transmit power reduction for both, mobile devices and BS. Therefore, it is crucial for the development of 5G wireless communications to have smart antenna designs that are optimized for directional gains, cost, and complexity.

The ultimate evolution to cell-free setups (in future systems) may initially require small-cell deployments. Indeed, deploying a large number of tiny cells, giving rise to heterogeneous networks (HetNets [35]), can help cope with the explosive growth of wireless traffic anticipated in the coming years. HetNets often include small, low-transmission-power cells in addition to the traditional macrocells. In such a dense 5G deployment, configuring and maintaining several servers and routers are challenging tasks that can be resolved by software-defined networks. This allows a division between the control and the data planes, bringing speed and flexibility to 5G networks, and may rebuild networks for fully automated management by stepping over open system interconnection (OSI) layers. Furthermore, cloud RAN aims to address the high data rate demands and promises to lower the cost of network deployment and operation while simultaneously enhancing system architecture, mobility, coverage performance, and energy efficiency [36].

The 5G standard was expanded by 3GPP Rel-16 [37] to include new verticals like (i) NR sidelink, which focuses on public safety and vehicle-to-everything; (ii) 5G broadcast, which is an evolution of LTE-based enhanced TV; (iii) nonpublic networks; (iv) Industrial IoT applications; and (v) shared or unlicensed spectrum, which includes sub-6 GHz bands. Additional changes in Rel-17 include multicast and broadcast services, expansion into FR2-2, support for satellite communications for smartphones and IoT devices, and the introduction of reduced capability (RedCap) devices aimed at new device types (e.g., wearables, surveillance cameras, and industrial sensors).

Although 5G NR has many technical components, it is important to note salient features such as scalable OFDM-based air interface, advanced channel coding (via low-density parity check coding, which provides more efficient support of throughput per decoder chip area), M-MIMO, and mm-wave operation.

5G networks have revolutionized spectrum management by incorporating a wide range of frequencies, from sub-6 GHz to millimeter-wave bands (24–52 GHz). This is critical for supporting the varied requirements of 5G services, including eMBB, URLLC, and mMTC. Spectrum sharing and dynamic spectrum access have become more prevalent, allowing for more flexible and efficient use of available resources. Techniques like dynamic spectrum sharing (DSS) enable 5G to coexist with 4G, providing a smoother transition and better spectrum utilization. Beamforming and M-MIMO technologies enhance the efficiency and capacity of the network by directing signals precisely where needed. Meanwhile, regulatory frameworks and global spectrum harmonization efforts have evolved to support these advanced techniques and allow a seamless operation of the 5G services worldwide.

The spectrum use evolution from 1G to 5G and key features of each generation are respectively illustrated and summarized in Figure 1.3 and Table 1.1.

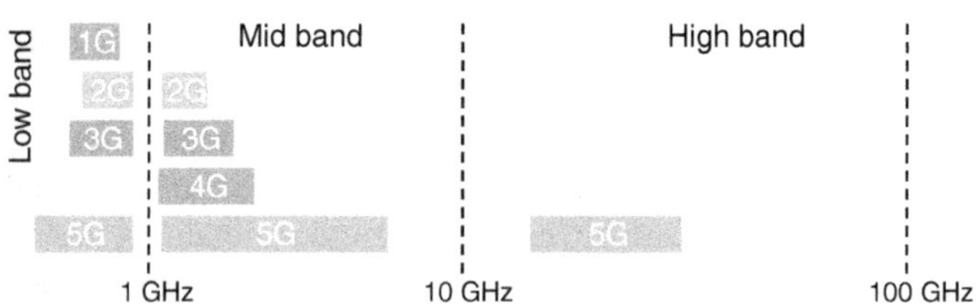

Figure 1.3 Spectrum utilization of different generations of cellular networks.

Table 1.1 Evolution of mobile networks.

Features	1G	2G	3G	4G	5G
Speed	2 kbps	64 kbps	2 Mbps	1 Gbps	> 1 Gbps
Latency	—	629 ms	212 ms	60–98 ms	< 1 ms
Access technology	AMPS, FDMA	GSM, TDMA, GPRS, EDGE	CDMA, UMTS, WCDMA, high-speed uplink packet access (HSUPA)/high-speed downlink packet access (HSDPA), EVDO	OFDMA/ SC-FDMA	BDMA, FBMC
Switching	Circuit	Circuit/Packet	Packet	Packet	Packet
Bandwidth	30 KHz	200 KHz	5 MHz	1.4–20 MHz	1–6 GHz
Forward error correction	NA	NA	Turbo codes	Turbo codes	Low density parity check codes
Application	Voice	Voice + data	Voice + data + video calling	Online gaming + high definition video	Ultra high definition video + virtual reality

Source: Adapted from [23, 38].

1.5 Beyond 5G and Prospects

After almost a decade of intense academic and industrial research on 5G mobile networks and the subsequent commercial deployment, it has become clear that 5G networks will fall short in supporting the vision of the Internet of Everything (IoE), i.e., the intelligent connection of people, processes, data, and things. Moreover, future networks are anticipated to connect many nodes in various proximity ranges, such as small cells, microcells, picocells, and femtocells, while there is an increasing need to support advanced bandwidth-hungry applications such as mixed reality, telepresence, and Industry 5.0 requiring infimum delay, highly reliable connections, and/or accessing the Internet via mm-wave/THz frequencies [39]. Given this, the research community has commenced intensive research on the Beyond 5G (B5G) successor technology, i.e., 6G mobile networks [39–41].

B5G networks are envisioned to overcome critical shortcomings of 5G mobile networks by adopting cell-free architectures and operating in sub-6 GHz, mm-wave, and THz spectrum. However, transforming such speculative ideas into real-time commercial deployments is a big challenge. Specifically, limited coverage is a key issue when operating in the mm-wave/THz spectrum and dedicated research is needed since current spectrum allocation methods are ineffective for such setups [42]. In this regard, Open RAN (O-RAN) may provide flexibility for spectrum occupancy in B5G mobile networks and thus facilitate spectrum management. Emerging open interface standards can be explored for minimizing interference between heterogeneous systems in a very congested B5G radio environment. In addition, O-RAN may permit operations in different spectrum bands, involving varied spectrum sharing (SS) levels over the mm-wave/THz spectrum. However, advanced SS techniques are required to deal with dynamically changing and highly heterogeneous environments.

Concerning M-MIMO, novel low-cost/complex hybrid architectures, and accompanied precoding techniques, are needed to promote sustainability and highly dense deployments [43]. In addition, artificial intelligence (AI) techniques can be used to predict channel characteristics and create dataset frameworks for beam-forming in M-MIMO [44]. Moreover, holographic displays are considered the next evolution in multimedia experience delivering 3D images from one or multiple sources to one or multiple destinations, providing an immersive 3D experience for the end user [45]. Interactive holographic capability in 6G networks will require a combination of very high data rates and ultralow latency. The key requirements will also include perfect synchronization, extreme resilience, and large computations. These requirements will serve as the backbone for advanced use cases such as Industry 5.0, which will need the expansion of digital twins so that visual models of the products, processes, and generation will allow better understanding and testing [46]. For Industry 5.0, 6G networks are expected to meet the intelligent information society standards to deliver ultrahigh reliability [47].

Mobile edge computing has been identified as the technology that will transform the 5G communication networks into distributed cloud computing platforms [48] and will continue its evolution over the next generations [49]. Mixed reality, autonomous driving, and holographic communications are all candidates for edge cloud coordination. The key network requirements include computing awareness of constituent edge facilities, joint network and computing resource scheduling, flexible addressing, and fast routing and rerouting.

The research community has already foreseen multiple techniques and applications that will leverage AI mechanisms in modern and future mobile networks [43, 50, 51]. For 6G networks, new network architectures comprising holistic network virtualization and pervasive AI will be required. The holistic network virtualization will consist of network slicing and digital twin to provision services and demands thereby, incorporating service- and user-centric networking [52]. The pervasive network intelligence will integrate AI into 6G networks from the perspectives of networking for AI and AI for networking. In turn, such an architecture will facilitate interplay between (i) digital twin and network slicing paradigms, (ii) model-driven and data-driven methods for network management, and (iii) virtualization and AI, to maximize flexibility, scalability, adaptivity, and intelligence for 6G networks.

Note that existing virtualization techniques mainly focus on service provisioning for communication networks [53]. On the other hand, virtualization, with a focus on service provisioning, will enable 6G to handle diverse coexisting services. However, this may not suffice for 6G since the characteristics of end user service demand will be the key to achieving user-centric networking. Hence, virtualization must focus on both, the service provision capability of a network and the service demand of end users in the network. Existing research on AI mostly addresses specific functions, layers, network segments, or applications of a network, while it is necessary to investigate the integration of AI into the complete network architecture. Indeed, a major drawback of the 5G technology is that it does not standardize the convergence of various functions, such as communication, sensing, security, computing, etc., at least natively [54]. Hence, 5G networks do not, and will not, realize the unlimited potential of AI [55], but may just use it to optimize the networks' legacy architecture and functions. This points toward the need to natively provide/exploit intelligence within 6G technology wherein, various 5G functionalities may converge for IoE and fully immersive experience provisioning.

AI technologies hold massive potential to improve network performance and minimize the overall network cost (both, CapEx and OpEx) [56]. Furthermore, they may solve link and system-level problems in 6G by including features such as self-configuration, aggregation, opportunistic setup, and context awareness [57]. As an example, it has been reported in [58] that an AI-enabled 6G network can realize radio signaling and maximum cognition to intelligent radio transmission. Notably, AI has already been attractive for implementation at various layers of 5G technology [59]. In addition, concerning spectrum resource management, most of the encountered algorithms are combinatorial, and thus cannot be implemented in practice. Therefore, AI-enabled algorithms will be mandated to outperform the current suboptimal approaches. These algorithms may suffice for performing the management of resource tasks with reduced supervision including, power control, beamforming in M-MIMO and cell-free environments, predictive scheduling, resource allocation, etc. In 6G networks, AI technology will reside in the new local cloud environments and will aid in creating multiple advanced applications via the use of sensors that have already been embedded within every aspect of human life.

Overall, the B5G research must target spectral-efficiency gains, high protection against interference, network congestion avoidance, high and massive mobility support, and resiliency and dependability [3, 60, 61]. In this regard, a broader and deeper understanding of link quality in multiple scenarios and a robust statistical taming of all sources of delay/error are immediate challenges [61, 62]. As a possible solution, the holistic integration of model and data-driven approaches and the synthesis of artificial data using generative adversarial networks can be conducted [61, 63]. Moreover, the big data resources in B5G networks will require high-security/privacy guarantees that may be provided with decentralized management, such as the one offered by Blockchain [64]. Meanwhile, specific to THz/mm-wave communication, the key challenge is that of severe path and absorption loss as mm-wave/THz carriers are unable to penetrate physical obstacles making them highly susceptible to blockage [65]. To address this issue, prediction techniques must be designed to infer the incoming blockage in beam direction [66]. Guaranteeing low power consumption and/or high energy efficiency will emerge as key challenges for B5G/6G devices, triggering further research on maturing and advancing backscattering communications [43, 67], wireless power transfer [68–70], tinyML [43, 71], energy-aware protocols [43, 72], wake-up radio [43, 73], metasurface-assisted communications [74, 75], to name a few.

Finally, the limited data amount from 5G spectrum assignment decisions and high uncertainty of market data for spectrum study is an issue under intensive research [76]. In addition, adapting network resources to meet the end user's QoS demands while simultaneously reducing the networks' power consumption is a must [77]. In addition, B5G networks must intelligently adapt to the changes in physical infrastructure, equipment, and data, under the regulatory framework set by policymakers [78].

B5G networks must further enhance spectrum management to support more demanding applications, such as holographic communications and Industry 5.0. For this, they will likely expand the operation to higher-frequency bands, including the THz spectrum, and require advanced spectrum sharing and interference mitigation strategies. O-RAN and AI-enabled spectrum management will be pivotal in dynamically adapting to the anticipated highly heterogeneous and dense environments. Specifically, O-RAN will allow operators to use open interfaces and virtualize network functions, enabling more efficient and flexible spectrum use. Meanwhile, AI can optimize spectrum allocation in real time, predicting traffic patterns and adjusting resources dynamically to minimize interference and maximize efficiency. The regulatory landscape will need to adapt to these advances. Future spectrum policies will likely focus on enabling more dynamic and flexible access models, including greater use of unlicensed and shared spectrum, to accommodate the exponential growth in wireless connectivity demands.

1.6 Conclusion

This chapter detailed the evolution of wireless mobile networks. We characterized every wireless network generation, including the specific aims, implemented technologies, and advantages/disadvantages. It is highlighted that a new mobile network generation is introduced almost every decade after the precedent, bringing significant KPI improvements, viz., data speeds, network capacity, latency, and reliability.

Spectrum management has been a crucial aspect throughout such an evolution. In 1G and 2G networks, spectrum allocation was relatively static, focusing on voice communication. The shift to 3G and 4G came with the introduction of spectrum auctions and dynamic spectrum access to meet the rising data demands. 5G further revolutionized spectrum management by incorporating a wide range of frequencies, enabling advanced techniques like dynamic spectrum sharing and M-MIMO to enhance the efficiency and capacity. In the chapter, we over-viewed key use cases offered by 5G networks and the progress toward future mobile networks. In fact, B5G networks are expected to adopt even more sophisticated spectrum management strategies, including the use of higher frequency bands such as THz, advanced spectrum sharing, and open radio access network (ORAN)- and AI-enabled management techniques. These advancements will support emerging applications such as holographic communications and Industry 5.0. We discussed prospects and key research avenues that require timely solutions.

Overall, understanding the progression of spectrum management is vital for the development and optimization of future wireless networks, ensuring they meet the increasing demands for connectivity and data services efficiently.

References

1 Matinmikko-Blue, M., Aalto, S., Asghar, M.I. et al. (eds.) (2020). White paper on 6G drivers and the UN SDGs. *6G Research Visions*, vol. 2020. http://urn.fi/urn:isbn:9789526226699 (accessed 30 September 2024).

2 Huang, Y., Jin, J., Lou, M. et al. (2022). 6G mobile network requirements and technical feasibility study. *China Communications* 19 (6): 123–136.

3 Mahmood, N., Lopez, O.L.A., Park, O. et al. (eds.) (2020). White paper on critical and massive machine type communication towards 6G [white paper]. *6G Research Visions*, vol. 11.

4 Farley, T. (2005). Mobile telephone history, *Privateline.com*. https://cdn6.f-cdn.com/files/download/438936/kaNGwTelenorPage_022-034.pdf.

5 Rappaport, T.S. (2010). *Wireless Communications: Principles and Practice*, 2e. Pearson Education India.

6 Abramson, N. (1970). The ALOHA system: another alternative for computer communications. *Proceedings of the 1970 Fall Joint Computer Conference*, volume 37, 281–285.

7 Walke, B.H. (2013). The roots of GPRS: the first system for mobile packet-based global internet access. *IEEE Wireless Communications* 20 (5): 12–23.

8 Hillebrand, F. (2013). The creation of standards for global mobile communication: GSM and UMTS standardization from 1982 to 2000. *IEEE Wireless Communications* 20 (5): 24–33.

9 Hata, M. and Nagatsu, T. (1980). Mobile location using signal strength measurements in a cellular system. *IEEE Transactions on Vehicular Technology* 29 (2): 245–252.

10 Gupta, A. and Jha, R.K. (2015). A survey of 5G network: architecture and emerging technologies. *IEEE Access* 3: 1206–1232.

11 Knisely, D.N., Kumar, S., Laha, S., and Nanda, S. (1998). Evolution of wireless data services: IS-95 to CDMA2000. *IEEE Communications Magazine* 36 (10): 140–149.

12 Adedoyin, M.A. and Falowo, O.E. (2020). Combination of ultra-dense networks and other 5G enabling technologies: a survey. *IEEE Access* 8: 22893–22932.

13 Telecommunications Industry Association (1993). Mobile station-base station compatibility standard for dual-mode wideband spread spectrum cellular system *TIA/EIA/IS-95*.

14 Ott, G.D. (1977). Vehicle location in cellular mobile radio systems. *IEEE Transactions on Vehicular Technology* 26 (1): 43–46.

15 Rappaport, T.S., Reed, J.H., and Woerner, B.D. (1996). Position location using wireless communications on highways of the future. *IEEE Communications Magazine* 34 (10): 33–41.

16 Reed, J.H., Krizman, K.J., Woerner, B.D., and Rappaport, T.S. (1998). An overview of the challenges and progress in meeting the E-911 requirement for location service. *IEEE Communications Magazine* 36 (4): 30–37.

17 Manninen, A.T. (2002). *Elaboration of NMT and GSM Standards: From Idea to Market*, no. 60. University of Jyväskylä.

18 Kebede, T., Wondie, Y., Steinbrunn, J. et al. (2022). Multi-carrier waveforms and multiple access strategies in wireless networks: performance, applications, and challenges. *IEEE Access* 10: 21120–21140.

19 Linge, N. and Sutton, A. (2015). *30 Years of Mobile Phones in the UK*. Amberley Publishing Limited.

20 Halonen, T., Romero, J., and Melero, J. (2004). *GSM, GPRS and EDGE Performance: Evolution Towards 3G/UMTS*. Wiley.

21 V.K. Garg, (2002). IMT-2000 GOAL. *Wireless Network Evolution: 2G to 3G*, Prentice Hall, 347.

22 Wijting, C., Doppler, K., Kalliojarvi, K. et al. (2009). Key technologies for IMT-advanced mobile communication systems. *IEEE Wireless Communications* 16 (3): 76–85.

23 Chen, S., Sun, S., Wang, Y. et al. (2015). A comprehensive survey of TDD-based mobile communication systems from TD-SCDMA 3G to TD-LTE (A) 4G and 5G directions. *China Communications* 12 (2): 40–60.

24 Abdullah, M.F.L. and Yonis, A. (2012). Performance of LTE release 8 and release 10 in wireless communications. *Proceedings Title: 2012 International Conference on Cyber Security, Cyber Warfare and Digital Forensic (CyberSec)*, 236–241. IEEE.

25 Sengupta, A., Alvarino, A.R., Catovic, A., and Casaccia, L. (2020). Cellular terrestrial broadcast—physical layer evolution from 3GPP release 9 to release 16. *IEEE Transactions on Broadcasting* 66 (2): 459–470.

26 Cao, J., Ma, M., Li, H. et al. (2013). A survey on security aspects for LTE and LTE-A networks. *IEEE Communications Surveys & Tutorials* 16 (1): 283–302.

27 Sankaran, C. (2012). Data offloading techniques in 3GPP Rel-10 networks: a tutorial. *IEEE Communications Magazine* 50 (6): 46–53.

28 Ghosh, A., Ratasuk, R., Mondal, B. et al. (2010). LTE-advanced: next-generation wireless broadband technology. *IEEE Wireless Communications* 17 (3): 10–22.

29 Holma, H., Toskala, A., and Reunanen, J. (2016). *LTE Small Cell Optimization: 3GPP Evolution to Release 13*. Wiley.

30 Ghosh, A., Maeder, A., Baker, M., and Chandramouli, D. (2019). 5G evolution: a view on 5G cellular technology beyond 3GPP release 15. *IEEE Access* 7: 127639–127651.

31 M Series (2015). IMT vision–framework and overall objectives of the future development of IMT for 2020 and beyond. *Recommendation ITU* 2083: 1–21.

32 GSMA Intelligence (2014). Understanding 5G: perspectives on future technological advancements in mobile *White paper*, 1–26.

33 You, L., Gao, X., Li, G.Y. et al. (2017). BDMA for millimeter-wave/terahertz massive MIMO transmission with per-beam synchronization. *IEEE Journal on Selected Areas in Communications* 35 (7): 1550–1563.

34 Wang, C.-X., Haider, F., Gao, X. et al. (2014). Cellular architecture and key technologies for 5G wireless communication networks. *IEEE Communications Magazine* 52 (2): 122–130.

35 Xu, Y., Gui, G., Gacanin, H., and Adachi, F. (2021). A survey on resource allocation for 5G heterogeneous networks: current research, future trends, and challenges. *IEEE Communications Surveys & Tutorials* 23 (2): 668–695.

36 Checko, A., Christiansen, H.L., Yan, Y. et al. (2015). Cloud RAN for mobile networks—a technology overview. *IEEE Communications Surveys & Tutorials* 17 (1): 405–426.

37 Baek, S., Kim, D., Tesanovic, M., and Agiwal, A. (2021). 3GPP new radio release 16: evolution of 5G for industrial Internet of Things. *IEEE Communications Magazine* 59 (1): 41–47.

38 del Peral-Rosado, J.A., Raulefs, R., López-Salcedo, J.A., and Seco-Granados, G. (2017). Survey of cellular mobile radio localization methods: from 1G to 5G. *IEEE Communications Surveys & Tutorials* 20 (2): 1124–1148.

39 Saad, W., Bennis, M., and Chen, M. (2020). A vision of 6G wireless systems: applications, trends, technologies, and open research problems. *IEEE Network* 34 (3): 134–142.

40 Iyer, S., Patil, A., Bhairanatti, S. et al. (2022). A survey on technological trends to enhance spectrum-efficiency in 6G communications. *Transactions of the Indian National Academy of Engineering* 7 (4): 1093–1120.

41 Mahmood, N.H., Alves, H., López, O.A. et al. (2020). Six key features of machine type communication in 6G. n *2nd 6G Wireless Summit (6G SUMMIT)*, 1–5.

42 Alsheyab, H.Y., Choudhury, S., Bedeer, E., and Ikki, S.S. (2019). Near-optimal resource allocation algorithms for 5G+ cellular networks. *IEEE Transactions on Vehicular Technology* 68 (7): 6578–6592.

43 López, O.L.A., Rosabal, O.M., Ruiz-Guirola, D.E. et al. (2023). Energy-sustainable IoT connectivity: vision, technological enablers, challenges, and future directions. *IEEE Open Journal of the Communications Society* 4: 2609–2666.

44 Bai, L., Wang, C.-X., Huang, J. et al. (2018). Predicting wireless mmWave massive MIMO channel characteristics using machine learning algorithms. *Wireless Communications and Mobile Computing* 2018: 1–12.

45 Petkova, R., Poulkov, V., Manolova, A., and Tonchev, K. (2022). Challenges in implementing low-latency holographic-type communication systems. *Sensors* 22 (24): 9617.

46 López, O.L.A., Porambage, P., and Liyanage, M. (2023). *Society 5.0 and Security*, 135–155.

47 Sedlmeir, J., Buhl, H.U., Fridgen, G., and Keller, R. (2020). The energy consumption of blockchain technology: beyond myth. *Business & Information Systems Engineering* 62 (6): 599–608.

48 Tomkos, I., Klonidis, D., Pikasis, E., and Theodoridis, S. (2020). Toward the 6G network era: opportunities and challenges. *IT Professional* 22 (1): 34–38.

49 Zhang, Y. and Zhang, Y. (2022). Mobile edge computing for beyond 5G/6G. *Mobile Edge Computing* 9 37–45.

50 Alsharif, M.H., Kelechi, A.H., Albreem, M.A. et al. (2020). Sixth generation (6G) wireless networks: vision, research activities, challenges and potential solutions. *Symmetry* 12 (4): 676.

51 Albreem, M.A., Alsharif, M.H., and Kim, S. (2020). A low complexity near-optimal iterative linear detector for massive MIMO in realistic radio channels of 5G communication systems. *Entropy* 22 (4): 388.

52 Shen, X., Gao, J., Wu, W. et al. (2022). Holistic network virtualization and pervasive network intelligence for 6G. *IEEE Communications Surveys & Tutorials* 24 (1): 1–30.

53 Zangiabady, M., Garcia-Robledo, A., Aguilar-Fuster, C., and Rubio-Loyola, J. (2020). A holistic framework for virtual network migration to enhance embedding ratios in network virtualization environments. *Journal of Network and Systems Management* 28 (3): 502–552.

54 David, K. and Berndt, H. (2018). 6G vision and requirements: is there any need for beyond 5G?. *IEEE Vehicular Technology Magazine* 13 (3): 72–80.

55 Zhou, F., Lu, G., Wen, M. et al. (2019). Dynamic spectrum management via machine learning: state of the art, taxonomy, challenges, and open research issues. *IEEE Network* 33 (4): 54–62.

56 Letaief, K.B., Chen, W., Shi, Y. et al. (2019). The roadmap to 6G: AI empowered wireless networks. *IEEE Communications Magazine* 57 (8): 84–90.

57 Sarieddeen, H., Saeed, N., Al-Naffouri, T.Y., and Alouini, M.-S. (2020). Next generation terahertz communications: a rendezvous of sensing, imaging, and localization. *IEEE Communications Magazine* 58 (5): 69–75.

58 Li, R., Zhao, Z., Zhou, X., Ding, G. et al. (2017). Intelligent 5G: when cellular networks meet artificial intelligence. *IEEE Wireless Communications* 24 (5): 175–183.

59 Piran, M.J. and Suh, D.Y. (2019). Learning-driven wireless communications, towards 6G. *International Conference on Computing, Electronics & Communications Engineering (iCCECE)*, 219–224.

60 Mahmood, N.H., Atzeni, I., Jorswieck, E.A., and López, O.L.A. (2023). Ultra-reliable low-latency communications: foundations, enablers, system design, and evolution towards 6G. *Foundations and Trends® in Communications and Information Theory* 20 (5-6): 512–747.

61 López, O.L.A., Mahmood, N.H., Shehab, M. et al. (2023). Statistical tools and methodologies for ultrareliable low-latency communication–a tutorial. *Proceedings of the IEEE* 111 (11): 1502–1543.

62 Cerar, G., Yetgin, H., Mohorcic, M., and Fortuna, C. (2021). Machine learning for wireless link quality estimation: a survey. *IEEE Communications Surveys & Tutorials* 23 (2): 696–728.

63 Creswell, A., White, T., Dumoulin, V. et al. (2018). Generative adversarial networks: an overview. *IEEE Signal Processing Magazine* 35 (1): 53–65.

64 Li, X., Jiang, P., Chen, T. et al. (2020). A survey on the security of blockchain systems. *Future Generation Computer Systems* 107: 841–853.

65 Restuccia, F. and Melodia, T. (2020). Deep learning at the physical layer: system challenges and applications to 5G and beyond. *IEEE Communications Magazine* 58 (10): 58–64.

66 Alrabeiah, M. and Alkhateeb, A. (2020). Deep learning for mmWave beam and blockage prediction using sub-6 GHz channels. *IEEE Transactions on Communications* 68 (9): 5504–5518.

67 Toro, U.S., Wu, K., and Leung, V.C.M. (2022). Backscatter wireless communications and sensing in green Internet of Things. *IEEE Transactions on Green Communications and Networking* 6 (1): 37–55.

68 López, O.L.A., Alves, H., Souza, R.D. et al. (2021). Massive wireless energy transfer: enabling sustainable IoT toward 6G era. *IEEE Internet of Things Journal* 8 (11): 8816–8835.

69 Alves, H. and Lopez, O.A. (2021). *Wireless RF Energy Transfer in the Massive IoT Era: Towards Sustainable Zero-energy Networks*. Wiley-IEEE Press.

70 Rosabal, O.M., López, O.L.A., Alves, H., and Latva-aho, M. (2023). Sustainable radio frequency wireless energy transfer for massive Internet of Things. *arXiv preprint arXiv:2306.15051*.

71 Kallimani, R., Pai, K., Raghuwanshi, P. et al. (2023). TinyML: Tools, applications, challenges, and future research directions. *arXiv preprint arXiv:2303.13569*.

72 Valente da Silva, R., López, O.L.A., and Souza, R.D. (2023). Energy-aware federated learning with distributed user sampling and multichannel ALOHA. *IEEE Communications Letters* 27 (10): 2867–2871.

73 Ruíz-Guirola, D.E., Rodríguez-López, C.A., Montejo-Sánchez, S. et al. (2022). Energy-efficient wake-up signalling for machine-type devices based on traffic-aware long short-term memory prediction. *IEEE Internet of Things Journal* 9 (21): 21620–21631.

74 Basharat, S., Hassan, S.A., Pervaiz, H. et al. (2021). Reconfigurable intelligent surfaces: potentials, applications, and challenges for 6G wireless networks. *IEEE Wireless Communications* 28 (6): 184–191.

75 Zhang, Z., Dai, L., Chen, X. et al. (2023). Active RIS vs. passive RIS: which will prevail in 6G? *IEEE Transactions on Communications* 71 (3): 1707–1725.

76 Matinmikko-Blue, M., Yrjölä, S., Seppänen, V. et al. (2019). Analysis of spectrum valuation elements for local 5G networks: case study of 3.5-GHz band. *IEEE Transactions on Cognitive Communications and Networking* 5 (3): 741–753.

77 López-Pérez, D., De Domenico, A., Piovesan, N. et al. (2022). A survey on 5G radio access network energy efficiency: massive MIMO, lean carrier design, sleep modes, and machine learning. *IEEE Communications Surveys & Tutorials* 24 (1): 653–697.

78 Latva-aho, M. and Leppänen, K. (2019). 6G white paper, key drivers and research challenges for 6G ubiquitous wireless intelligence. *University of Oulu, Tech. Rep. Vers*, vol. 1.

2

Spectrum Access Options for Local 6G Networks

Marja Matinmikko-Blue[1], Seppo Yrjölä[2], and Petri Ahokangas[3]

[1] *Infotech Oulu and Centre for Wireless Communications, University of Oulu, Oulu, Finland*
[2] *Centre for Wireless Communications, University of Oulu, Nokia, Oulu, Finland*
[3] *Oulu Business School, University of Oulu, Oulu, Finland*

2.1 Introduction

Local mobile communication networks have become an emerging deployment model in the 5G era following important early attempts in the 4G era [1]. Nationwide cellular mobile communication network deployments by mobile network operators (MNOs) are started to be complemented by local deployments of cellular networks by a variety of stakeholders in several countries. These local 4G and 5G networks often aim at serving the local needs of different user groups, and their establishment depends on local spectrum availability.

Spectrum authorization decisions defining who can use the spectrum, where and how, play a key role in the local mobile communication business. Recently, 5G spectrum authorization decisions have introduced local spectrum licensing schemes in several countries [2], where the national regulator issues local spectrum access rights directly to different stakeholders for deploying their own cellular mobile communication networks [3]. These spectrum access models based on local licensing have allowed different stakeholders to apply for local spectrum access rights, opening the traditionally closed mobile communication market to a variety of stakeholders. In addition to local spectrum licenses issued by the regulator, there are other mechanisms to access the precious radio spectrum for deploying local mobile communication networks as analyzed, e.g., in [3, 4]. These options include, e.g., leasing of spectrum bands assigned to MNOs as well as unlicensed access to spectrum. The spectrum access options have different levels of spectrum sharing, different stakeholders involved, and different spectrum valuation and pricing approaches. This results in a complex

Intelligent Spectrum Management: Towards 6G, First Edition.
Edited by Sridhar Iyer, Anshuman Kalla, Onel Alcaraz López, and Chamitha De Alwis.

environment for stakeholders to choose between the options to access the radio spectrum for local mobile communication networks.

6G as the next generation of mobile communication networks is expected to continue the 5G path of nationwide networks complemented with local cellular network deployments by a variety of stakeholders including MNOs and others. Spectrum for 6G is expected to continue to utilize a mix of different frequency bands as in mobile communication systems but with potentially larger bandwidths and higher operating frequencies [5]. Different bands could be jointly utilized to provide wireless links that satisfy a wide range of use cases' requirements [5].

To help 6G research and development in academia, industry, and policy-making, this chapter aims to bring clarity to how different stakeholders could gain access to the radio spectrum for local mobile communication networks. Specifically, the chapter aims to answer the following research questions:

- What kind of spectrum access options could exist for local 6G networks?
- How could the identified spectrum access options be evaluated from spectrum management, sharing, valuation, and pricing perspectives?

The rest of the chapter is organized as follows. Section 2.2 presents the state of the art on spectrum management in the context of local mobile communication networks. Section 2.3 dwells deeper into spectrum valuation and pricing. Section 2.3.1 identifies key spectrum access options for local 6G networks. Section 2.4 analyses the presented spectrum access options considering spectrum management, sharing, valuation, and pricing perspectives. Finally, conclusions are drawn in Section 2.5.

2.2 Background/State of the art

This section summarizes the state of the art on local mobile communication networks and related spectrum management approaches.

2.2.1 Local Mobile Communication Networks

Local mobile communication networks became an emerging deployment model in the 5G era following early attempts in the 4G era. Prior work introduced "local 5G micro-operator" concept for local vertical-specific service delivery using local 5G networks [1]. The aim was to open the 5G mobile communication market to allow different stakeholders to establish their own local mobile communication networks with different levels of MNO involvement. A major challenge was how to efficiently manage timely access to spectrum for the deployment of the local 5G networks.

In general, the end users of the local mobile communication networks can be closed user groups, resulting in private networks, or open user groups that can become public networks, depending on the specifics of the deployment and country [2]. Examples of such networks include stand-alone nonpublic network (SNPN) for closed user groups or a public network integrated nonpublic network (PNI-NPN) for open user groups, such as MNOs' customers. In addition, a neutral host model is possible, where the local network is serving the end users of another service provider or several service providers, including either a closed user group or an open user group. The local mobile communication networks can ultimately serve a mix of closed and open user groups. Depending on the end user groups served, the local networks may be subject to different regulatory approaches in different countries.

The deployment of private 5G networks in industrial locations for robots and machines resulting in closed end user groups has gained increasing attraction. Local mobile communication networks target at exploiting new business opportunities to serve the long-tail of various vertical sectors' needs, which would allow different agile local players to deploy and operate the networks. Specific examples include transport hubs like harbors and airports and industrial sites like factories and mines. Operations in these examples are restricted to a specific geographical area, which can have a high demand for communications and makes a local network an option. On the other hand, local mobile communication networks can also be deployed by the traditional MNOs leveraging existing network infrastructure, spectrum licenses, and capabilities in planning, deploying, managing, and optimizing mobile networks efficiently while offering interoperability and mobility with adjacent public networks.

Most recently, local 6G networks have gained attention as a deployment model allowing different stakeholders to deploy future 6G networks [6]. 6G will combine communication with other services including, e.g., sensing, imaging, positioning, and computing. In local 6G networks, this can result in offering location-specific services that make use of different resource combinations and knowledge of the specific operational environment aiming at cost savings and/or new business opportunities. It is expected that the demand for access to local spectrum will be more dynamic in the future, supporting more temporary use of spectrum.

2.2.2 Spectrum Management for Local 5G Networks

2.2.2.1 Spectrum Management Approaches

Radio spectrum is the fundamental resource for all wireless communications, which makes it a key topic for consideration in 5G including their local deployments. The radio spectrum is managed at different levels including international, regional, and national (see, e.g., [7]) resulting in complex stakeholder interactions aiming at efficient and effective use of the scarce natural resource realized by technical solutions. Spectrum management approaches can be broadly divided into three categories: administrative allocation, market-based mechanisms, and the unlicensed commons model [3], which are discussed in more detail in the following in the context of local 5G networks.

Administrative allocation is a form of hierarchical command and control approach, where the regulator defines spectrum property rights and awards them to applicants based on some criteria. Administrative allocation used to dominate as the regulators' common mechanism to assign spectrum access rights over a long time period [3, 8]. Administrative allocation typically uses beauty contests, first come first served, or other types of direct awards, where the regulator assigns spectrum access rights to applicants. Administrative allocation is still in use for many radio communication services by many national regulatory authorities, for example, for noncommercial governmental services. Administrative allocation typically aims to create rules and conditions for spectrum access that minimize harmful interference between the different wireless systems. Especially, the protection of incumbent spectrum users' rights is of interest in this approach by defining the spectrum property rights in a manner that avoids harmful interference. Applicants typically need to show how they advance the public interest in order to gain spectrum access rights. Administrative allocation is applicable as a rather simple mechanism, when demand does not exceed supply. It has received growing criticism over time in competitive environments where more applicants wish to gain spectrum access rights.

Market-based mechanisms have replaced administrative allocation in many countries in assigning access rights, especially for mobile communications starting in the 3G era [9]. Market-based mechanisms determine spectrum property rights and grant licenses to use the radio spectrum to a limited number of applicants through some market mechanisms. The most common mechanism is spectrum auction [10]. Market-based mechanisms take the value of spectrum into account and incentivize more efficient spectrum use. Access rights awarded through market-based mechanisms often allow secondary markets for spectrum, where the original spectrum property rights can be traded or leased in whole or in part by geography, bandwidth, or by both for a given duration. Spectrum auctions are a common way to award spectrum access rights to deploy mobile communication networks from the 3G era onward. Auctions also involve spectrum and interference management work from the National Regulatory Authority (NRA).

Unlicensed commons approach allows a larger number of wireless systems to access the radio spectrum without defining spectrum property rights [11]. The most typical approach is to allow unlicensed devices to operate under given regulator-defined rules and conditions, often for low-power wireless devices. The unlicensed commons approach allows market entry to a large number of users without having to wait for license awarding decisions. Unlicensed commons approach has been considered a major promoter of innovation and competition, currently carrying large amounts of data wirelessly locally.

The above-described three general spectrum management approaches can be and have been used in different frequency bands. For example, spectrum bands for 5G in Europe include low bands (700 MHz), mid-bands (3.5 GHz), and high bands (26 GHz bands). These bands have different characteristics in terms of, e.g., frequency division duplex or time division duplex (FDD/TDD) techniques with different synchronization patterns, different cell radii, bandwidth capacity, and outdoor to indoor propagation feasibility, which in turn directly influence the feasible deployment models. The spectrum awarding decisions taken by the regulators in the new 5G bands have varied significantly between different countries. In Europe, the European Commission has mandated to study the band 3.8–4.2 MHz as harmonized band for local mobile communication networks [12].

2.2.2.2 The Role of Spectrum Sharing

Spectrum sharing in general refers to the situation, where two or more radio systems operate in the same frequency band [13]. There are many definitions for spectrum sharing and they vary depending on the purpose leading to conflicting stakeholder views on spectrum sharing. In general, there are two main categories of spectrum sharing including horizontal and vertical spectrum sharing. Horizontal spectrum sharing refers to the situation where different radio systems are operating in the same band with same level of spectrum access rights [13]. Vertical spectrum sharing refers to the situation where the radio systems have different levels of spectrum access rights [13]. In realistic spectrum-sharing situations, both horizontal and vertical spectrum sharing are typically present, see, e.g., Ref. [14]. Examples include the commercialized US-based Citizens Broadband Radio Service (CBRS) and the standardized but not deployed European Licensed Shared Access (LSA). CBRS includes vertical spectrum sharing between three tiers of spectrum access rights, which are based on three distinct spectrum management approaches. It also includes horizontal spectrum sharing between radio systems in the same tier of spectrum access rights. LSA introduces additional licensed users in bands that already have authorized usage [15]. The LSA concept includes vertical spectrum sharing where entrant licensed users are allowed to access spectrum bands, which are used by incumbent spectrum users who are protected from harmful interference to a predefined degree [16]. The LSA concept also includes horizontal spectrum sharing when more than one radio system is admitted in to the band as the entrant.

MNOs have traditionally obtained exclusive spectrum access rights over large areas, leaving spectrum sharing to be an internal matter. With 5G, techniques to enable 5G and 4G to share the same band have emerged (so-called dynamic spectrum sharing), which is an example of intraoperator horizontal spectrum sharing, see Ref. [17]. In some bands used for 5G, there is vertical spectrum sharing to protect the incumbent spectrum users of the band, such as in the 26 GHz band in some countries, which is an example of intersystem vertical spectrum sharing. Spectrum sharing using techniques to manage resulting interference scenarios is an important enabler for local mobile communication networks including techniques for power control, signal detection, frequency selection, beamforming, and use of databases among others. Local spectrum licensing adopted in several countries is an example of horizontal spectrum sharing, where the same spectrum can be used by different stakeholders to deploy networks with a geographical separation distance. The separation distance can be shortened by techniques to manage the interference.

2.2.2.3 Spectrum Access Options for Vertical Service Providers

Vertical service providers have emerged aiming to serve the needs of vertical sectors (see, e.g., [4]). Prior work on spectrum access options for local mobile communication networks in [4] has identified five options for vertical 5G service providers:

1. **Unlicensed**: No license required by the regulators (e.g., industrial, scientific, and medical [ISM] bands).
2. **Secondary Licensing**: The National Regulatory Authority (NRA) issues a license to the vertical network service provider (VNSP) in a band already used by an incumbent spectrum user (e.g., LSA and CBRS).
3. **Spectrum Trading or Leasing**: A license holder such as an MNO trades or leases spectrum to the VNSP.
4. **Virtual Network Provided by MNO**: MNO provides a dedicated virtual network for the VNSP.
5. **Local Licensing**: NRA issues a local license to the VNSP.

The identified spectrum access options were analyzed in [4] with respect to reliability, locality and mobility, MNO involvement, time scale issues, costs, and investment security attributes. Additionally, spectrum access options for local indoor network deployments by local operators were assessed in [18], where the local operators were either collaborating closely with the MNOs or deploying their own independent networks. Regarding the first option (unlicensed), globally available 5 GHz unlicensed spectrum is widely used by vertical service providers. Figure 2.1 shows unlicensed 5 GHz band availability globally. Additionally, regarding the first option (unlicensed), a new 6 GHz unlicensed spectrum for RLAN and 5G NR-U networks is available as shown in Figure 2.2.

Regarding the second option (secondary licensing), there are developments that have made assigned but unused spectrum available for other stakeholders.

Figure 2.1 Unlicensed 5 GHz spectrum availability.

Figure 2.2 New 6 GHz unlicensed spectrum for RLAN and 5G NR-U networks.

For example, Ofcom has introduced a "shared access license," which allows bands assigned to MNOs to be used by another licensee if not used by the original holder of spectrum access rights.

Regarding the third option (spectrum trading or leasing), spectrum transfer and leasing have been possible in European Union (EU) member countries for a

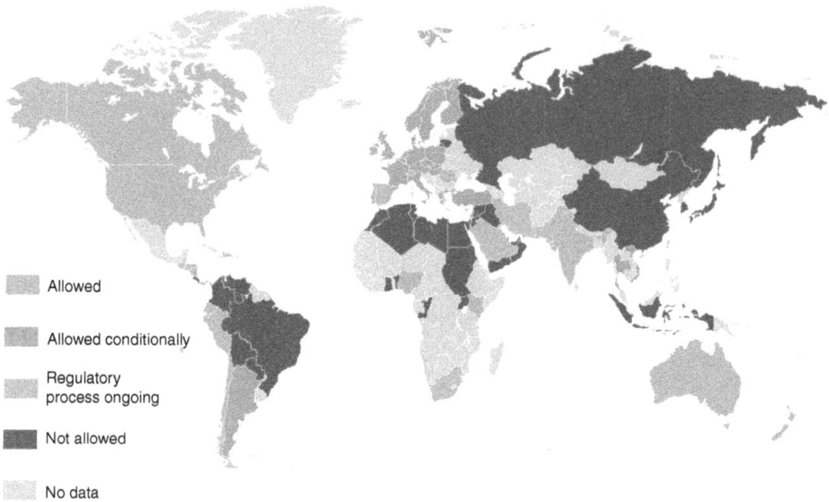

Figure 2.3 Spectrum trading and leasing availability.

long time, promoting spectrum transactions [19]. However, national approaches can vary. Figure 2.3 presents the status of spectrum trading and leasing globally. Conditional allowance means that each proposed lease needs to be approved by the national regulator, which usually aims to ensure that there is no harmful impact on competition.

The fourth option (virtual network provided by MNO) is not a specific spectrum access option as such but more like a collaboration model between the MNO and the vertical service provider. In this model, the MNO provides a virtual network for the vertical service provider using the MNOs spectrum resources. This model is specifically relevant in countries where spectrum sub-leasing is not allowed.

Regarding the fifth option (local licensing), several countries have introduced local licensing including Finland, Germany, the United Kingdom, Japan, etc.; see Refs. [6, 20]. In Europe, several countries have started to issue local licenses in 5G spectrum awarding, but the spectrum bands and approaches are versatile, see Ref. [20] for an overview. Figure 2.4 illustrates the status of locally licensed spectrum for 4G and 5G.

2.3 Spectrum Valuation and Pricing

Next, we consider spectrum valuation and pricing, which are important factors in the allocation of spectrum to different radio communication systems and

Figure 2.4 Spectrum trading and leasing availability.

assignment of rights of use of spectrum to different stakeholders to deploy wireless systems. Spectrum valuation and pricing mechanisms cover approaches to reach balances between spectrum demand and supply across the radiocommunication systems and stakeholders.

2.3.1 Spectrum Valuation in Mobile Communications

The general goals for spectrum management include effectiveness by allocating spectrum to the right usage, and efficiency by assigning it to those who value it the most [21]. Spectrum management decisions aim to maximize the value of the spectrum, its efficient utilization, and its benefits to society [21]. These high-level goals highlight the importance of the value of spectrum. Spectrum valuation is a complex process with multiple perspectives that are emphasized differently in various countries [11, 22–24]. Moreover, different wireless services, such as mobile broadband communications, private mobile radio, broadcast, satellite, and military use, have a different basis for their value due to their distinct business models, technologies, and role in society. Ultimately, the value of spectrum can be determined via its relative worth, utility, or importance to a stakeholder and is therefore contextual (related to use), defined by demand or need, relational, and may evolve over time [25–27]. Indeed, value exchange [28, 29] for the sake of value accumulation can be seen as the key driver for any economic activity and it is substantiated by an array of different value processes, including value (co)-creation, delivery, (co)-capture, and sharing.

Recent discussions have started to broaden the generic meaning of value to include customer needs, economic return, and compliance, and societal and environmental goals to ensure sustainable value creation [30]. The price of spectrum, in turn, refers to its value in monetary units or other compensation when it is offered, bought, or sold. In general, price can be based on cost, competition in the markets, of customer value [27]. Recent discussions on pricing have addressed its dynamic nature as related to uncertainties and risks in the market environment. The problem or conflict of determining the value and price of spectrum can be approached from the perspective of the principal-agent theory [31] for explaining differing priorities or conflicts in asymmetrical hierarchical and contractual relationships between the principal as asset owner and the agent as asset controller to whom the control or use of the asset has been delegated [32]. In the case of spectrum access policymaking contexts such set ups occur including a regulator and an MNO, or an MNO and a local operator. The conflict between the principal and agent can become worse due to lack of information/transparency regarding the markets or activities between the stakeholders or due to weak or undefined/underdefined control and coordination mechanisms.

The mobile communications market typically involves a small number of MNOs that have gained long-term exclusive spectrum licenses from the regulator. Spectrum auctions have become the dominant mechanism for MNOs to gain spectrum access rights. In some countries, spectrum auctions have resulted in significant income for the governments while other countries have had different goals than collecting money, indicating different views on the value of spectrum. The impact of mobile communication on society goes beyond spectrum auction revenues and is a complicated topic to analyze [9, 10, 33, 34]. Competition can generate greater governmental revenues in the long term compared to the auction revenues [10], which further complicates the topic.

The allocation of local spectrum licenses for establishing local mobile communication networks to different stakeholders in addition to traditional country-wide licenses to MNOs has opened the topic of how to value and price the local licenses [3]. A functioning marketplace for trading local spectrum access rights will require a cost-effective and accurate method for valuating and pricing the licenses. There has been limited research on how the value of the spectrum licenses is determined or distributed on a local level. The demand and value of local spectrum licenses can drastically change between the different locations and types of use, influencing the pricing of local licenses. The commercial value of spectrum is the price of a private 4G/5G radio spectrum license, should it be for sale, and can be determined by the expected future cash flows that the license generates for its holder. These cash flows are generated through additional increases in revenues or reductions in costs. The license can, for example, be used to offer a new product to increase revenues, or to implement a new manufacturing method to reduce costs. Licenses can also be used defensively to limit competition. By blocking competition, the license holders can use their improved market power to increase cash flows. Additionally, Marks et al. [35] note that radio spectrum licenses hold a significant option value. Typically, general spectrum pricing models used by regulators consist of the following variables: (i) the opportunity cost for a given band and location, (ii) the amount of spectrum used, (iii) the type of service, (iv) the frequency band, and (v) the location [35].

The pricing of spectrum licenses needs to consider the above-mentioned underlying economic factors. Conventionally, the price of a license for mobile communications is determined through spectrum auctions, whose rules are defined for the given market. Given enough buyers, auctions result in relatively accurate pricing as the buyers can use sophisticated, context-specific financial models to estimate the economic factors. However, due to the small value and illiquidity of licenses for local 4G and 5G networks, auctions are often not a viable method of pricing for local licensing. One local license can, e.g., only cover the area of a single factory or port, which means that the license has only one potential buyer. A small number of buyers makes auctions inefficient [36]. Additionally, the context-specific financial models used by the buyers in auctions, as well as in the bidding process, can be very resource-intensive. These costs can be relatively large compared to the commercial value generated by the trade of a local license for private 4G and 5G networks.

Cramton and Doyle [37] state that an open access market for spectrum would increase competition and make the process more efficient, transparent, fair, and simple. On the other hand, there are several problems related to the spectrum market. In designing any markets, there are four general causes of market failures [38]: (i) information asymmetry, (ii) externalities spill-over costs or benefits accrue to anyone not involved in a given interaction, (iii) monopoly power, and (iv) commercial and technical risks. Particularly related to market liquidity in the secondary spectrum access market, Chapin and Lehr [39] have found three framing characteristics: (i) low availability due to low achievable Quality of Services (QoS) and hence, low demand; (ii) high transaction costs; and (iii) risks related to technical interference. Furthermore, Xavier and Ypsilanti [40] have highlighted the following issues related to the introduction of secondary markets: (i) uncertainty regarding the future primary allocations leading to incorrect estimations of spectrum scarcity and value (option value); (ii) lack of information on available spectrum; (iii) risks of increased interference; (iv) coordination, harmonization, and controlling mechanisms; (v) anti-competitive conduct, in specific concentration of spectrum and hoarding; disruptive effects on end users; and (vi) ability to achieve public interest objectives.

Berry et al. [41] further state that the current secondary spectrum exchange is inefficient because of regulatory friction and complexity-driven transaction costs. Transaction costs can be divided into three broad categories [42]: (i) search and information costs, which are costs related to determining that the required spectrum (in quality, quantity, terms, price, etc.) is available on the market; (ii) bargaining costs, which are costs required to come to an acceptable agreement, drawing up an appropriate business and regulatory contracts (on asset markets, it is some function of the distance between the bid and ask; and (iii) policing and enforcement costs, which are the costs of making sure the other party follows the terms of the contract in business, technology, and policy, and taking appropriate action often through the regulatory and legal system if this turns out not to be the case. Factors influencing resource pricing are depicted in Figure 2.5.

In the current proposals of secondary markets, the valuation of the spectrum licenses is, in general, proposed to be done by spectrum auctions. However, Tonmukayakul and Weiss [36] state that auctions work only when the licenses are liquid, i.e., when there are enough demanders (buyers) and suppliers (sellers). It is reasonable to assume that the number of buyers per specific case is relatively small, e.g., in the vertical and enterprise licensing cases, because the licenses are local, and they benefit only few buyers. For example, in a case where a factory or a harbor wants to deploy a local network to its own property, there are no other buyers because the property is only used and/or controlled by the facility owner.

In many countries, transfer of spectrum via spectrum trading or leasing is possible and even on the regional service area level, like in the United States, Canada, Australia, and India. However, it is often hard to gauge what can really be accomplished in practice as several above-discussed causes of market failure exist, and there is a lack of clarity about the rules and nature of transfers that are allowed. There is also a basic assumption, e.g., in Europe that spectrum would be transferred on a national basis between "traditional" MNOs rather than at specific locations to a private network provider. Furthermore, trading is complicated by

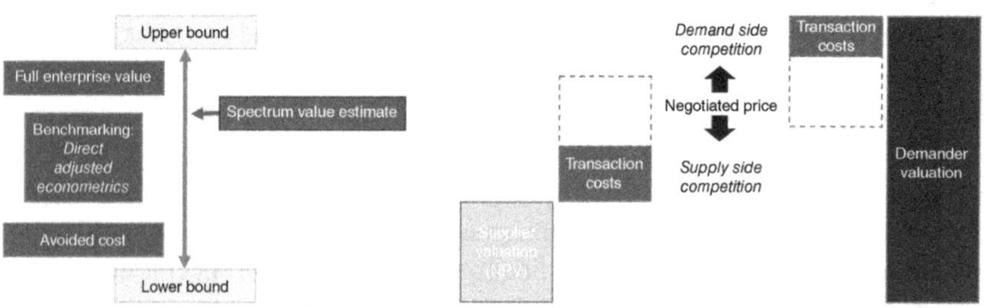

(a) (b)

Figure 2.5 Valuation methods (a) and factors influencing resource prices (b).

the fact that public mobile and private networks have very different use-cases, which makes the transfer of a public mobile license with its corresponding license conditions and obligations to a private network use highly impractical. Leasing is a more realistic option, but the appropriate regulatory framework and financial incentives need to be present for this to happen. The economics of providing radio services affects users' interest in buying, selling, or leasing spectrum. There is an information asymmetry problem between regulators and operators, and substantial differences in business cases among operators and potential new entrants.

The research in [35] has concluded that the value of spectrum is different for MNOs and local operators as well as dependent on the business cases: the spectrum supplier (typically an MNO) must be incentivized to share or lease spectrum, and the demander (e.g., a private network operator) must maximize the value realized from using the spectrum. From the technical operations perspective, there can be a lack of complementarity between spectrum use and the underlying radio infrastructure resulting in a sunk costs problem. In many cases, spectrum cannot readily be released and used without changing equipment. The costs of making these changes can be significant relative to the value of the spectrum and hence act as an impediment to frequent trading of the spectrum. Moreover, harmonization of spectrum bands for uses and associated equipment standardization reduces the potential to move spectrum to a higher value use. Only in very large markets are spectrum licensees able to offer equipment vendors the scale required to produce low-cost equipment—in these cases large increases in spectrum value may be realized through trades.

Flexibility for license holders, i.e., MNOs in most cases, to set the terms of the leasing agreement and transaction costs, which include search and information, bargaining, and policing and enforcement, are highly dependent on the NRA. From the strategic perspective, market structure and strategic behavior may reduce willingness to trade or lease spectrum. Most mobile markets comprise just a handful of market players who often view spectrum as a strategic asset that enhances their competitive positions in the market, and further they may consider option cost arising from uncertainty related to no longer having access to the spectrum asset too high.

An alternative to administrative allocation and the market-based mechanisms is the commons approach that puts spectrum access into the hands of many, by allowing unlicensed access under regulator-defined rules and conditions, e.g., for low-power wireless devices [11, 43]. The value of unlicensed spectrum under the commons approach arises from making spectrum access possible to several diverse types of usages and stakeholders, which has resulted in new profits, as well as from its capability to promote innovation and competition. The commons approach also allows faster time to market, as entrants do not need to wait for license awarding decisions. However, unlicensed spectrum has been a sufficient enabler and solution for best-effort service, and it does not result in the guaranteed quality required by many vertical applications.

2.3.2 Spectrum Valuation Methods

Commonly used spectrum valuation methods can be categorized into three groups: (i) benchmarking, (ii) avoided cost modeling, and (iii) full enterprise value as shown in Figure 2.5. Direct benchmarking [44] is a historical data-based approach using values for a frequency band or band group (e.g., 1, 2, and 3 GHz ranges) as benchmarks for the same band. In cases where there are few actual data points in a frequency band-adjusted benchmarking is implemented based on the use of past international benchmarks as reference values, adjusted with local observed ratios of selected reference bands relative to the band to be valued. The benchmarking method is a statistical econometric analysis that provides an objective way of controlling for market and economic factors and is less reliant on comparisons between different countries. Instead, it uses the estimated relationships between observed spectrum prices and the explanatory factors to predict a value based on the economic and demographic circumstances in the country and location under study. The goal is to estimate the value of spectrum based on the effects of various drivers of spectrum value using data from a large number of auctions. Potential drivers of spectrum values and their variables are, e.g., spectrum characteristics, spectrum allocation and assignment/license characteristics, auction characteristics, and economic and market characteristics.

Avoided cost modeling [18] is particularly functional in the cost-based valuation of the supply-side operator spectrum resource. This engineering value is based on the calculation of the potential cost reduction that an operator could expect if an additional spectrum is available in its spectrum portfolio following an award process. The cost savings arise from the possibility of using the extra spectrum as a means of enhancing capacity or extending coverage without incurring the additional costs of physical infrastructure rollout. This is analogous to the opportunity cost of the spectrum from the perspective of its use for the delivery of public mobile services. Avoided cost modeling requires country and location-specific context and operator-specific information, such as demand estimates, technical network dimensioning parameters, and network costing that represent actual network operating conditions.

Full enterprise valuation [21, 45] modeling assesses the value of spectrum from a commercial perspective estimating the profit the spectrum in question will generate for an operator over the license term. In this model, the net present value (NPV) of the spectrum to an operator is calculated. Estimating the value of the spectrum involves the analysis of the impact on profits because of changes in spectrum fees over the modeling period. A discounted cash flow (DCF) model used involves two components: total revenue estimated based on current and future demand (subscribers/users and average revenue per user), market circumstances (competition), and network costs (radio access network and core network), non-network expenses (selling, general, administrative expenses, and tax). Modeling requires a significant amount of information such as operators' business costs, market revenues, and future business plans and strategies, which will involve technology and market projections over the span of the spectrum license. Assumptions utilize information from other markets that are estimated based on existing sources, and views on competition, investment decisions, and market and regulatory environments will need to be agreed on. As the demand side value that the private network operator will realize is uncertain, it is not feasible to set MNO spectrum valuation at a profit-maximizing level. Instead, it is useful to consider the minimum price that may result in a transaction. This minimum price will be at a level that incentivizes the MNOs to share their spectrum and is dependent on the conditions of sharing itself. Additionally, an optional approach that expands NPV with an options value, depicted in Figure 2.5, may be used in uncertain cases where flexibility is needed in how and when spectrum is used, e.g., when a spectrum license is tradeable, renewable, or fungible in terms of a change in use. The option value of spectrum is the value related to having the flexibility to invest at the optimal time; where there are irreversible costs associated with making investments, there is the possibility of waiting for the latest information to arrive and potentially avoiding or reducing uncertainty over investment returns. Method relies on running scenario analytics on business plans and network rollout plans as the sum of additional equipment costs incurred due to not having access to spectrum, and any revenues foregone due to spectrum being unavailable. Option value sets the strategic value [23] to the spectrum that may occasionally exceed the basic full enterprise valuation.

2.4 Analysis of Identified Spectrum Access Options for Local 6G Networks

Building on the prior work on spectrum access for local 5G networks summarized in Section 2.2 and the ongoing development of future 6G networks, in this section we identify spectrum access options for the deployment of local 6G networks. We then analyze the spectrum access options with respect to the elements identified in Section 2.2 including spectrum management model and sharing, as well as valuation and pricing discussed in Section 2.3. We note that MNOs can deploy their own local 6G networks using their existing and/or upcoming spectrum assets. Thus, the case of spectrum access for the MNO in its own spectrum holdings is omitted here. Instead, we focus on the situation of gaining access to spectrum for local 6G networks beyond MNO deployments in their own spectrum assets allowing different stakeholders to establish the local 6G networks. The resulting spectrum access options are available for the MNOs as well. As a result, we classify the resulting spectrum access options to deploy local 6G networks into (i) local spectrum licenses from the NRA, (ii) local spectrum access rights acquired from incumbent spectrum user(s), (iii) local spectrum access rights acquired from a third party (new brokerage model), and (iv) unlicensed access. All of them are discussed in more detail in the following.

We note that MNOs can deploy their own local 6G networks using their existing and/or upcoming spectrum assets. Thus, the case of spectrum access for the MNO in its own spectrum holdings is omitted here. Instead, we focus on the situation of gaining access to spectrum for local 6G networks beyond MNO deployments in their own spectrum assets allowing different stakeholders to establish the local 6G networks. The resulting spectrum access options are available for the MNOs as well. As a result, we classify the resulting spectrum access options to deploy local 6G networks into the following four main categories, which are discussed in more detail in this section:

1. Local spectrum licenses from the NRA,
2. Local spectrum access rights acquired from incumbent spectrum user(s),
3. Local spectrum access rights acquired from a third party (new brokerage model), and
4. Unlicensed access.Key aspects of identified spectrum access options are discussed in more detail in the following.

2.4.1 Local Licenses from the NRA

The first considered spectrum access option to establish local 6G networks by different stakeholders is to obtain a local spectrum license directly from the NRA. In this model, the NRA defines spectrum property rights and awards the rights of use. In this situation, the NRA defines the rules and conditions of accessing the band and issues the licenses for the stakeholder to deploy a local 6G network. The contractual relationship takes place between the NRA as the principal and the local licensee as the agent. Regarding spectrum valuation and pricing, the NRA determines and collects the fee for the local licenses from the licensees. Table 2.1 summarizes pricing examples from different countries for local licenses from the NRA.

Local licenses from the NRA in 5G spectrum awards have become a reality in several countries as summarized in Section 2.2, but the variety between countries is high in terms of which bands, if any, are made available through local

Table 2.1 The pricing examples of the locally licensed spectrum in the mid-bands below 6 GHz.

Country	Description	Spectrum pricing (annual fee)
Australia	3.4–4.0 GHz band area-wide apparatus licensing in the minimum allocated bandwidth of 10 MHz. The smallest geographic area is a single cell comprising an area of approximately 1.8 km × 1.8 km	AU$/MHz/pop price × bandwidth (MHz) × population of geographic area
Canada	3.65–3.7 GHz wireless broadband services (WBS) band light licensing (all-come, all-served) in Tier 4 Service areas	Depends on the population of the service area, fees range from 250 Canadian dollar (CAD) in rural areas up to 21 kCAD in the Service Area of Toronto
Czech Republic	3.4–3.8 GHz band licensed to operators with regulatory commitment to lease spectrum in the range of for the purpose of Industry 4.0. nonpublic electronic communications network. Area related to land ownership or usage right	Price defined by regulator: (i) One-off price for establishment of a service is Czech koruna (CZK) 50,000 (excl. VAT); (ii) Annual price for a service (CZK): $m/12 \times 1000 \times V \times a$, where m is the number of months of lease per calendar year; V is the leased width of a spectrum range in MHz; and a is the actual surface area of the land estate in km^2
Denmark	3.74–3.8 GHz spectrum licensee has a regulatory obligation (during the first four years) to lease spectrum to other entities, for private networks	The max price has been determined by the regulator (Danish krone [DKK]): (Fee + Rp) × bandwidth (BW) × Area/Country area, where Fee is the fee per MHz paid by lessor to Danish Energy Agency (DEA) (5640 DKK in 2020); Rp is the price per MHz paid in auction last round for obligation free 3.5 GHz spectrum (251545 DKK); BW is the leased amount (MHz); Area is the lease area + additional 500 m protection distance; and Country area is 43,123 km^2
Finland	2.30–2.32 GHz band is available to local 4G/5G networks on a licensed basis	Depends on bandwidth, and also population coverage in case it exceeds 275,000. Examples: 5 MHz → The Euro (EUR) 279.83, 10 MHz → EUR 559.66, 20 MHz → EUR 1119.31

(Continued)

Table 2.1 (Continued)

Country	Description	Spectrum pricing (annual fee)
Germany	3.7–3.8 GHz band licensed for local and regional private network deployments, for industrial applications. Access by MNOs is currently allowed. Eligibility for application: owner of the property, tenant, lessee, or someone commissioned by the owner	Fee in EUR is equal to $1000 + B \times t \times 5 \times (6\,a1 + a2)$, where EUR 1000 is the base amount; B is the bandwidth in MHz (min. 10 to max. 100 MHz); t is the duration of the allocation in years (e.g., 10 years); a is the area in km^2 with a differentiation between the settlement and traffic area ($a1$) and other areas ($a2$)
Japan	4.6–4.9 band for private 5G local deployments	The license fee for a base station (BS) is 5900 yen/year, and for a mobile station is 370 yen/year
France	2575–2615 MHz band available for local enterprise deployments. BW < 20 MHz, unless strong arguments for more. The license can be spread over several locations (and time) for one licensee	Annual Fee (in EUR) is $L \times Bf \times c \times a \times k6$, where L is the bandwidth in MHz (10 up to 40 MHz); Bf is a function of the frequency band (=8.7); c is a function of coverage area (e.g., 0.0005 for areas between 1 and 5 km^2); a is a function of the type of authorization by band (=2); and $k6$ is a value linked to the propagation characteristics of the band (=34, 000). Example: for 20 MHz, 1 year, and 5 km^2, the fee is 20 MHz × 8.7 × 0.0005 × 2 × 34,000 = EUR 5912 per year
Lesotho	2.52–2.62 GHz spectrum for private networks using FDD or TDD configuration	Application fee M10000 (ca EUR 550). Annual fee for access spectrum M11800 per MHz (ca EUR 650 per MHz)
Norway	3.8–4.2 GHz band is available for local private mobile networks. The low-power license is within a radius of 50 m from a preapproved center location. Several low-power permits can be granted to ensure coverage in a limited geographical area. The high-power license for outdoor use is to cover larger areas based on the approval of a single BS	Annual license prices (low power/high power): 20 MHz: Norwegian Krone (NOK) 200/1000; 40 MHz: NOK 400/1400; 60 MHz: NOK 800/2200; 80 MHz: NOK 1600/3800

Netherlands	3.7–3.8 GHz band available for local private LTE networks in southern parts of the country. License is required, per BS or per BS sector. One license for several femto- or pico-cells	The license fee is EUR 696, and the annual fee is EUR 332 per license
South Korea	4.72–4.82 GHz band for private/local 5G service. The license can be requested by the premise (land or building) owner, tenant, or trusted 3rd party. License period of two to five years to be chosen by applicant. Spectrum License (Local operators) or BS license (own network) required	Total cost = unit price $\times (5a1 + a2 + 1) \times$ period (years) $\times$ *BW* (blocks), where the unit price is South Korean Won (KRW) 100K (EUR 75)/10 MHz/year. Coverage in km^2 is specified through using $a1$ in big cities with a population over 500,000 people or $a2$ in other cities, e.g., 100 MHz in a big city costs KRW 6M. Free of charge for nonprofit cooperation. 4500/*per* km^2/year
Sweden	3.72–3.8 GHz band is available on a regional basis to support industrial 5G and IoT services. The minimum frequency block is 10 MHz, with steps of 10 MHz to a maximum of 40 MHz. Licensing is based on applications, and coverage is related to properties or parts of properties	Annual fee of Swedish krona (SEK) 524 (about EUR 50) per 10 MHz (SEK 500 is the fee for licenses and SEK 24 is the fee for market control)
The United Kingdom	3.8–4.2 GHz band available under a Shared Access license on a first-come first-served licensing basis. Low power licenses (per area, $r = 50$ m) and medium power licenses (per BS), normally issued only for rural areas	Annual license fee Great British Pound (GBP) 80 per 10 MHz.

licensing and how the licenses are awarded. Regarding the spectrum management model, gaining access to spectrum for local 6G networks through local licenses from the NRA can take place under either the administrative allocation or the market-based mechanism spectrum management model. Local spectrum licenses could be acquired from the NRA in different types of spectrum bands including:

- spectrum bands that are cleared from incumbent use for local licensing by the NRA to deploy local 6G networks.
- spectrum bands that are in use by MNOs, where additional local licenses are awarded by the NRA to deploy local 6G networks.
- spectrum bands that are in use by other incumbent spectrum users than MNOs, where additional local licenses are awarded by the NRA to deploy local 6G networks.

Regarding spectrum sharing, local licensing from the NRA involves horizontal spectrum sharing between local license holders. Typically, there is a geographical separation distance between licenses in the same frequency channel. The separation distance can be made shorter with interference mitigation techniques. The degree to which vertical spectrum sharing exists depends on the bands' existing usage. In spectrum bands that are cleared from incumbent spectrum use before local licenses are awarded, there is no vertical spectrum but only horizontal spectrum sharing between the local licensees. In spectrum bands that are in use by MNOs and additional local licenses are awarded to deploy local 6G networks, there is vertical spectrum sharing between the MNO and the local license holder. In spectrum bands that are in use by other type of incumbent spectrum users than the MNOs having higher levels of spectrum access rights, the awarding of additional local licenses to deploy local 6G networks with lower level of spectrum access rights results in a more complex spectrum sharing situation and interference scenarios. In that case, there is vertical spectrum sharing between the incumbent(s) and the local licensees as well as horizontal spectrum sharing between the local license holders calling for more sophisticated interference mitigation approaches and techniques. The latter two spectrum bands are examples of where the LSA concept is applicable [15, 16], where additional authorized users are introduced in bands with existing users while protecting a certain level of guarantee for both the incumbents and the entrants.

2.4.2 Local Spectrum Access Rights Acquired from Incumbent Spectrum User(s)

The second considered spectrum access option to establish local 6G networks is to acquire rights to use the spectrum from the incumbent spectrum user(s), who have the spectrum access rights. Spectrum access rights could be acquired from different types of incumbents, such as MNOs and satellite operators, depending on the national conditions. The difference to the first considered spectrum access option is that the current holder of spectrum access rights transfers the rights of use to the stakeholder wishing to deploy a local 6G network instead of the NRA issuing spectrum access rights. In this model, the contractual relationship takes place between the incumbent spectrum user as the principal and the local licensee as the agent. Regarding spectrum valuation and pricing, the price of spectrum is primarily determined by the incumbent (or by NRA if the incumbent is now allowed to determine the price). The pricing is impacted by the incumbent's own need for using the spectrum, competition situation, as well as strategic position on whether to open the spectrum for use by others. In the case where local spectrum access rights are acquired from an MNO, which currently holds spectrum access rights in the geographical area of interest, the MNO may have different approaches to defining the rules and conditions as well as pricing for accessing the band. This is allowed in many countries, but implementations are less common. Additionally, the NRA could play a role, depending on national situations. Local spectrum access rights could be leased from the MNO in different types of spectrum bands:

- existing bands, where the MNO holds spectrum access rights, or
- possible new spectrum bands for 6G awarded to MNOs.

Regarding the spectrum management model, gaining access to spectrum for local 6G networks through spectrum leased from incumbent spectrum user(s) typically falls under the market-based mechanism. In the case of an MNO being the incumbent and leasing its spectrum to local use, they agree on the conditions. The NRA may have a say, or its approval may be needed, depending on national conditions.

Regarding spectrum sharing, in existing spectrum bands, where the MNO holds spectrum access rights, there is vertical spectrum sharing between the MNO and the local network. In possible new spectrum bands for 6G awarded to MNOs, there is similarly vertical spectrum sharing between the MNO and the local network. Horizontal spectrum sharing is involved between the local networks.

2.4.3 New Brokerage Models

The third considered spectrum access option to establish a local 6G network is to acquire spectrum access rights from a third party, which here is called a spectrum broker. In this model, the third party (broker) does not own the rights of use of the spectrum but transfers them between the existing holder of rights and the entrant wishing to get the rights. In this situation, the broker defines the rules and conditions for accessing the band in collaboration with those whose rights are transferred. Typically, this type of operation and role is not common, and its availability depends on the national conditions. In this model, the contractual relationship takes place between the incumbent and the local licensee through the broker. The incumbent remains the principal asset owner and in charge of the spectrum use, and the broker becomes the agent as asset controller to whom the control or use of the asset has been delegated and acts as the principal. The local licensee is the final agent. One challenge in the brokerage model is that spectrum is typically assigned for certain kinds of use and the change of the type of use can be difficult. The NRA's role is important in order to make spectrum transfers efficient and feasible.

The brokerage model, where local spectrum access rights are procured from a third party, could fall under the market-based mechanism. Regarding spectrum valuation and pricing, the value depends on the broker's business model as well as the original holder of spectrum access rights. The spectrum value can be based on potential benefits from new usage and transaction fees from transferring of rights. Broker's compensation could be a percentage of the spectrum transaction. The brokerage model would require a tool for creating the spectrum marketplace to efficiently manage the transactions and control the potential interferences between different systems in order to match the different needs in demand and supply side as well as to support enough transactions.

The brokerage spectrum access model would bring together spectrum assets from different frequency bands with potentially different levels of spectrum access rights. The existing holder of spectrum access rights could offer some spectrum assets for brokering while at the same time maintaining some level of access rights, depending on the contract being made. The brokerage model could be established on different types of spectrum bands:

- spectrum bands cleared from incumbent spectrum users
- spectrum bands with existing incumbent spectrum users

Regarding spectrum sharing, in spectrum bands that are cleared from incumbent spectrum users, there will be horizontal spectrum sharing between the local networks. In spectrum bands with existing incumbent spectrum users, there will be vertical spectrum sharing between the incumbents and the entrants and horizontal spectrum sharing between the entrants.

2.4.4 Unlicensed Access

The fourth considered spectrum access option for local 6G networks is through unlicensed access. Gaining access to spectrum for local 6G networks through unlicensed access falls under the unlicensed commons spectrum management model. In this model, the NRA acts as the principal and local 6G networks are the agents together with other systems operating in the unlicensed band. There can be a large number of users of different types accessing the same unlicensed band. Regarding spectrum valuation and pricing, the traditional approach has been to not collect fees but allow access for free. It is possible to collect fees (e.g., administrative fees) to complement any costs that may arise from more efficient technical mechanisms for managing interference between users.

Unlicensed access to spectrum bands could take place in different types of spectrum bands including (see, e.g., [46]):

- existing unlicensed bands (e.g., 2.4, 5, 6, and 60 GHz)
- possible new unlicensed bands in the 6G era

Regarding spectrum sharing, in existing unlicensed bands (e.g., 2.4, 5, 6, and 60 GHz), the local 6G networks need to share the spectrum with other users according to the rules of the bands. Horizontal spectrum sharing takes place between unlicensed users through a variety of mechanisms. Vertical spectrum sharing takes place when there are incumbents to be protected such as in the 5 GHz unlicensed spectrum. In the possible new unlicensed bands in the 6G era, new rules for spectrum sharing need to be developed.

2.5 Conclusion

This chapter has identified four general categories of spectrum access options for different stakeholders to establish local 6G networks. The MNOs can deploy local mobile communication networks using their existing and upcoming spectrum holdings. It must be noted that in this chapter the local 6G deployments by the MNOs are omitted. The focus has been on presenting a variety of possible local spectrum access options in the 6G era to inspire further research considering technical, business, and/or regulatory perspectives about the different local spectrum access options.

The identified four spectrum access options include local licenses from the regulator, spectrum acquired from the incumbent spectrum users, new spectrum brokerage model, and unlicensed access. They have different characteristics in terms of spectrum management models, spectrum sharing, and spectrum valuation and pricing. All these options require further research in terms of different techniques for managing the resulting interference scenarios, potential spectrum bands for local 6G networks, suitable spectrum awarding models, and spectrum pricing and valuation methods, which are in the infancy for local networks. The future spectrum access models will probably be a mix of those introduced in this chapter.

Acknowledgment

The authors would like to acknowledge Heidi Himmanen from Traficom and Oxana Gisca from Oulu Business School. In addition, funding from Business Finland in 6G-Bridge local 6G project and the Research Council of Finland (former Academy of Finland) in 6G Flagship Programme (Grant No.: 346208).

References

1 Matinmikko, M., Latva-aho, M., Ahokangas, P. et al. (2017). Micro operators to boost local service delivery in 5G. *Wireless Personal Communications* 95 (1): 69–82.

2 Matinmikko, M., Latva-aho, M., Ahokangas, P., and Seppänen, V. (2018). On regulations for 5G: micro licensing for locally operated networks. *Telecommunications Policy* 42 (8): 622–635.

3 Matinmikko-Blue, M., Yrjölä, S., Seppänen, V. et al. (2019). Analysis of spectrum valuation elements for local 5G networks: case study of 3.5-GHz band. *IEEE Transactions on Cognitive Communications and Networking* 5 (3): 741–753.

4 Vuojala, H., Mustonen, M., Chen, X. et al. (2020). Spectrum access options for vertical network service providers in 5G *Telecommunications Policy* 44 (4): 101903.

5 ITUR Sector (2022). Future technology trends of terrestrial international mobile telecommunications systems towards 2030 and beyond. ITU-R, Tech. Rep.

6 Matinmikko-Blue, M., Yrjölä, S., and Ahokangas, P. (2023). Spectrum management for local mobile communication networks. *IEEE Communications Magazine* 61 (7): 60–66.

7 Lehtilä, S., Alén, A., Korpisaari, P., and Himmanen, H. (2023). Spectrum regulation and frequency allocation in the context of a smart city–using the regulatory approach in Finland as an example. *Information and Communications Technology Law* 32 (3): 418–432.

8 Hazlett, T.W. (2008). Optimal abolition of FCC spectrum allocation. *Journal of Economic Perspectives* 22: 103–128.

9 Klemperer, P. (2002). What really matters in auction design. *Journal of Economic Perspectives* 16 (1): 169–189.

10 Cramton, P. (2013). Spectrum auction design. *Review of Industrial Organization* 42 (2): 161–190.

11 Bazelon, C. (2009). Licensed or unlicensed: the economic considerations in incremental spectrum allocations. *IEEE Communications Magazine* 47 (3): 110–116.

12 European Commission (2021). Mandate to CEPT on technical conditions regarding the shared use of the 3.8-4.2 GHz frequency band for terrestrial wireless broadband systems providing local-area network connectivity in the union. EC, Tech. Rep.

13 ITUR Sector (2014). Cognitive radio systems (CRSs) in the land mobile service. ITU-R, Tech. Rep.

14 A. Patil, S. Iyer, O.L.A. López, R.J. Pandya, K. Pai, A. Kalla, and R. Kallimani, “A Comprehensive Survey on Spectrum Sharing Techniques for 5G/B5G Intelligent Wireless Networks: Opportunities, Challenges and Future Research Directions”. Computer Networks, Elsevier, vol. 253, 2024.

15 Electronic Communications Committee (ECC) within the European Conference of Postal and Telecommunications Administrations (CEPT) (2014). Licensed Shared Access. ECC Report 205,” ECC CEPT, Tech. Rep.

16 Matinmikko, M., Okkonen, H., Palola, M. et al. (2014). Spectrum sharing using licensed shared access: the concept and its workflow for LTE-advanced networks. *IEEE Wireless Communications* 21 (2): 72–79.

17 Iyer, S. (2022). Performance analysis of a dynamic spectrum assignment technique for 6G. *IETE Journal of Research* 69 (11): 7695–7703.

18 Ahmed, A.A.W., Markendahl, J., and Ghanbari, A. (2013). Evaluation of spectrum access options for indoor mobile network deployment. *IEEE 24th International Symposium on Personal, Indoor and Mobile Radio Communications (PIMRC Workshops)*, London.

19 Electronic Communications Committee (ECC) within the European Conference of Postal and Telecommunications Administrations (2011). Description of Practices Relative to Trading of Spectrum Rights of Use. ECC Report 169. ECC CEPT, Tech. Rep.

20 Radio Spectrum Policy Group (2023). Radio Spectrum Policy Group Draft Opinion on the Development of 6G and Possible Implications for Spectrum Needs and Guidance on the Rollout of Future Wireless Broadband Networks. RSPG23-026, Tech. Rep.

21 Beltran, F. (2017). Accelerating the introduction of spectrum sharing using market-based mechanisms. *IEEE Communications Standards Magazine* 1 (3): 66–72.

22 Bazelon, C. and McHenry, G. (2013). Spectrum value. *Telecommunications Policy* 37: 737–747.

23 Molleryd, B.G., Markendahl, J., Makitalo, O., and Werding, J. (2010). Mobile broadband expansion calls for more spectrum or base stations-analysis of the value of spectrum and the role of spectrum aggregation. *1st European Regional ITS Conference*, Copenhagen, Denmark

24 ITD Sector (2012). Exploring the value and economic valuation of spectrum. ITU-D, Tech. Rep.

25 Anderson, J.C., Jain, D.C., and Chintagunta, P.K. (1992). Customer value assessment in business markets: a state-of-practice study. *Journal of Business-to-Business Marketing* 1 (1): 3–29.

26 Töytäri, P., Rajala, R., and Alejandro, T.B. (2015). Organizational and institutional barriers to value-based pricing in industrial relationships. *Industrial Marketing Management* 47: 53–64.

27 Iveroth, E., Westelius, A., Petri, C.J. et al. (2013). How to differentiate by price: proposal for a five-dimensional model. *European Management Journal* 31 (2): 109–123.

28 Wilson, D.T. (2003). Value exchange as the foundation stone of relationship marketing. *Marketing Theory* 3 (1): 175–177.

29 Verstraete, T. and Jouison-Laffitte, E. (2011). *A Business Model for Entrepreneurship*. Cheltenham: Edward Elgar.

30 Bocken, N.M., Rana, P., and Short, S.W. (2015). Value mapping for sustainable business thinking. *Journal of Industrial and Production Engineering* 32 (1): 67–81.

31 Waterman, R.W. and Meier, K.J. (1998). Principal-agent models: an expansion? *Journal of Public Administration Research and Theory* 8 (2): 173–202.

32 Kim, H. (2023). Bridging principal-agent and mechanism design theories: an integrated conceptual framework for policy evaluation. *Asia Pacific Education Review* 25 (2): 329–342.

33 Hazlett, T.W. and Munoz, R.E. (2009). A welfare analysis of spectrum allocation policies. *The RAND Journal of Economics* 40 (3): 424–454.

34 Cave, M. and Nicholls, R. (2017). The use of spectrum auctions to attain multiple objectives: policy implications. *Telecommunications Policy* 41 (5-6): 367–378.

35 Marks, P., Pearson, K., Williamson, B. et al. (2009). Stimating the Commercial Trading Value of Spectrum. *Report for Ofcom*. Plum Consulting.

36 Tonmukayakul, A. and Weiss, M.B. (2008). A study of secondary spectrum use using agent-based computational economics. *Netnomics* 9 (2): 125–151.

37 Cramton, P. and Doyle, L. (2015). An open access wireless market. White Paper.

38 Roth, A.E. (2007). The art of designing markets. *HBR 85, no 10.*

39 Chapin, J.M. and Lehr, W.H. (2007). Cognitive radios for dynamic spectrum access-the path to market success for dynamic spectrum access technology. *IEEE Communications Magazine* 45 (5): 96–103.

40 Xavier, P. and Ypsilanti, D. (2006). Policy issues in spectrum trading. *INFO* 8 (2): 34–61.

41 Berry, R., Honig, M.L., and Vohra, R. (2010). Spectrum markets: motivation, challenges, and implications. *IEEE Communications Magazine* 48 (11): 146–155.

42 Dahlman, C.J. (1979). The problem of externality. *Journal of Law and Economics* 22 (1): 141–162.
43 Carter, K.R. (2006). Policy lessons from personal communications services: licensed vs. unlicensed spectrum access. *CommLaw Conspectus* 15: 93–117.
44 Shaw, B.A., Beltran, H.F., and Sowerby, K.W. (2017). Valuing spectrum at mm wavelengths for cellular networks. *15th International Telecommunications Society (ITS) Acia-Pacific Regional Conference* (24–27 June 2017). Kyoto, Japan.
45 Cave, M. and Pratt, N. (2016). Taking account of service externalities when spectrum is allocated and assigned. *Telecommunications Policy* 40 (10-11): 971–981.
46 Iyer, S. (2023). Machine learning enabled dynamic spectrum access for 6G wireless networks. *Journal of Applied Security Research* 19 (2): 330–350.

Section II

3

Spectrum Management Technologies in Mobile Networks

Harri Saarnisaari

Centre for Wireless Communications, University of Oulu, Oulu, Finland

3.1 Background

Spectrum is needed to transmit data and control information from a base station to a user equipment and vice versa. The former direction is called a downlink (DL) direction and the latter an uplink (UL) direction. The DL and UL directions could be separated in time or frequency domains using time division duplex (TDD) and frequency division duplex (FDD) techniques, respectively. Usually, a base station serves more than one user at a time. The resources for each user can be arranged in time, frequency, or both, and recently also in space within a base station cell. The first arrangement is called time division multiple access (TDMA), the second one frequency division multiple access (FDMA), their hybrid as TDMA/FDMA or the fourth-generation (4G) system form as orthogonal frequency division multiple access (OFDMA), and the last one space division multiple access (SDMA), which is combined with one or more other techniques. It is a task of the scheduler to allocate time, frequency, and space resources to users, in both DL and UL direction. Additional multiple access schemes are carrier sensing multiple access (CSMA) and code division multiple access (CDMA). In the former, user devices as well as access points sense if the channel was free before they require access, and in the latter, users' data are put in the same time-frequency element but separated with spreading codes. A base station may have one carrier frequency per cell or more than just one. In the latter case, user data is allocated into one carrier or, if carrier aggregation was applied, several carriers.

These spectrum usage alternatives of various generations of mobile cellular systems are shown in Table 3.1. The system capacity within a cell depends on allocated bandwidth, the number of carriers, and signal-in-space characteristics such

Intelligent Spectrum Management: Towards 6G, First Edition.
Edited by Sridhar Iyer, Anshuman Kalla, Onel Alcaraz López, and Chamitha De Alwis.

Table 3.1 Mobile cellular system features.

System	DL/UL division (dB)	Multiple access	Several carriers
2G	FDD	TDMA/FDMA	Yes
3G	FDD	CDMA	Yes
4G	FDD/TDD	OFDMA	Yes
5G	FDD/TDD	OFDMA/SDMA	Yes
WiFi	CSMA	CSMA	Yes

as modulation and coding. A part of frequency usage is frequency hopping, where a user is changing frequency within a cell to counteract frequency fading. In the second-generation (2G) system, the user data is jumped from a carrier frequency to another, but in 4G and the fifth-generation (5G) system, user data is jumped within the instantaneous bandwidth.

So far, different ways in which spectrum is used in cellular systems is discussed. It is time to move to address spectrum management, i.e., carrier frequency allocations. Before doing this, it is reminded that spectrum may also be needed for a backhaul (or transport) link, i.e., between the base station and the point where the core network can be achieved, if they were implemented using wireless radio-frequency technology. Historically, access and backhaul have had dedicated allocated frequencies (e.g., microwave links and satellite links), but integrated terrestrial and nonterrestrial solutions may call for more advanced solutions.

3.2 Cell Frequency Planning

Cell frequency planning is the core business in the spectrum management. It has been so since the beginning of cellular radio systems. It means how an operator is sharing its allocated bandwidth among cells such that these do not interfere with each other but, on the other hand, maximize the total throughput of operator's system.

A basic principle is that the same frequency cannot be used in cells covering the same geographical area. Indeed, there must be a safety margin as illustrated in Figure 3.1 to protect UL and DL transmissions also at cell edges. Consequently, a frequency reuse factor is significantly determining operator's system capacity. Actually, next-generation cellular systems (e.g., 5G after 4G) are first installed in places that need more capacity since they provide new spectrum and therefore, more capacity. At the same time, new generations are offering higher data rates

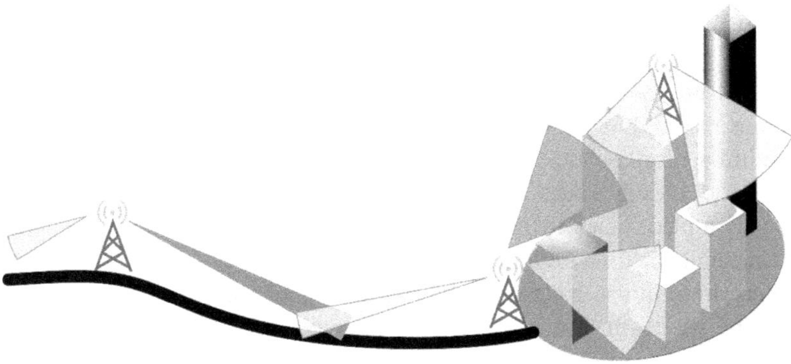

Figure 3.1 An illustration of frequency planning including frequency reuse issue. The sector colors denote different carrier frequencies. Overlapping cells should use different carrier frequency.

and may make old generations obsolete. This is in many places happening to the third-generation (3G) system since 4G is offering adequate Internet connectivity that 3G could not do, and coverage areas are the same. At the moment, it seems that 5G is not replacing 4G in that way.

Historically, base stations have covered outdoor areas sectorially. This means that a base station along a road has two sectors, one in each direction. In cities, a base station often has 360° coverage, e.g., using four 90° sectors. These are usually called macro base stations covering a large geographical area. If the transmitted power was reduced and/or higher carrier frequencies used, coverage area is limited. However, in order to cover the same area as using a macro base station, these base stations should be populated more densely than macro base stations. This would make frequency planning (management) more challenging. Low-power indoor base stations have appeared in the portfolios of base stations to handle indoor coverage problems. Therein, interference is not such a problem since walls, etc. limit signal propagation at used high frequencies. Anyway, WiFi is a very popular indoor coverage technology.

A base station coverage area is affected by an antenna radiation pattern (in three dimensions of which the up-dimension has not been very important so far) and transmission power (effective isotropic radiated power [EIRP] actually). A lot of efforts have been put in cell planning tools that typically use radio propagation models predicting the cell coverage area. A network designer can affect antenna height, antenna orientation including tilting, transmit power, and carrier frequency. A way to boost the cell capacity is carrier aggregation, meaning that more than one carrier is used in the cell. In addition, spectrum management unrelated means have been developed for boosting the capacity. These include co-operative transmission between base stations in the area.

3.3 Steps Toward Dynamic Spectrum Access

Today, spectrum is a scarce natural resource since traditionally a chunk of spectrum is reserved for certain purposes like cellular mobile radio services and then, within a chunk, allocated to mobile network operators (MNOs). In many countries, allocation has been nationwide. This static allocation means that even if regionally a piece of spectrum was not often used (underused) or was totally unused, it cannot be used for other purposes.

A better way to utilize spectrum is known as dynamic spectrum access (DSA). Therein, the spectrum is used on a need basis but such that all systems get fair share. Different rules and algorithms are needed and have been developed. There could be both intersystem DSA for collaboration between MNOs (or systems) and intrasystem DSA to handle frequency allocation within an MNO or system. The goals of these two could be a bit different. Other interesting DSA feature is its tempo. The tempo could be slow or fast. A slow tempo occurs if frequency changes are needed very frequently, e.g., when a system is installed and then again maybe a few years later when other systems appear in the area. Still this is DSA since frequency allocation is done automatically, not manually. At the other end of tempo, there are systems that actively look for free spectrum to operate, where decisions are made in second or even subsecond level.

Full-fledged, autonomous DSA methods have so far been too much for practice. Instead, limited versions have been adopted in the form of frequency sharing. Licensed shared access (LSA) and citizens broadband radio service (CBRS) have been adopted in Europe and the United States, respectively. LSA is standardized by European Telecommunications Standards Institute (ETSI) and CBRS by the US Federal Communications Commission (FCC). Both protect the incumbent radio system in the considered frequency band, but when the incumbent system is not active in an area, they allow other systems to utilize the band. The incumbent system denotes the system to whom the frequency band is allocated by the national frequency regulation body. LSA and CBRS are using different rules and are at the moment used just on certain frequency bands. The main idea is to allow MNOs and other users, such as public protection disaster relief (PPDR) and program making and special events (PMSE) users, to use unused or underutilized frequencies locally. The use may be short or long duration.

The use of these technologies is slowly spreading. Some experiments have been made and first users have appeared.

3.3.1 LSA

LSA is defined in ETSI technical specification [1] and its evolved version in [2]. The LSA system consists of an LSA repository and an LSA controller. The involved entities are a national frequency regulation body, an incumbent user, or the one that wishes to license the spectrum.

The repository keeps incumbent's usage information and protection requirements. It is connected to the LSA controllers and shares availability information with them and receives acknowledgment information from them. It allows the regulation body to monitor the operation of the LSA system and provides the LSA system with information on the sharing framework and the LSA licensees. Finally, the repository ensures that the LSA system operates in conformance with the sharing framework and the licensing regime. The controller is located at the LSA licensee's domain. In addition to above-stated information sharing, it controls radio parameters such that the usage is according to the agreed license.

3.3.2 CBRS

CBRS is a three-tier system. The first tier includes incumbent users such as military radars. The second tier consists of priority licensee users for which a part of used spectrum is allocated. The third tier is for general access that may use the whole band if it is (locally) unused.

CBRS system consists of a network of sensors detecting the use of the CBRS band and a spectrum access system (SAS). Radio system operators that want to locally use the CBRS band ask SAS to reserve unused channels. They will pay a SAS fee for spectrum and may charge their customers to cover the fee and get profit.

3.3.3 TVWS

Television (TV) broadcast signals at a certain frequency are not necessarily covering the whole nation, leaving some areas underused. TV white space (TVWS) technology aims to use this underused spectrum locally.

National frequency regulation bodies describe if TVWS is allowed or not within a nation and rules for usage if it is allowed. Typically, TV coverage maps are needed (for each frequency) as well as radio parameter control for TVWS users such that they do not interfere with TV broadcast signal in an adjacent area. The usage may be per the regulation body license that is somewhat rigid way or more flexible using an LSA or a CBRS type system.

3.3.4 Summary So Far

The primitive DSA systems discussed so far have common features. First, the systems have to know local spectrum usage status that may be called local spectrum situation awareness. It could be based on a database, measurements, or both. Second, a user has to ask permission and can enter only if access is granted. A system entity is corresponding to share these permissions.

Temporal changes occur very slow, i.e., systems are not suited for quickly occurring changes in system load and thus spectrum need. National frequency regulation bodies are involved in the process and new improvements may therefore have very slow progress. These systems have not been a success story so far, though the use of CBRS is slowly growing.

3.3.5 5G NR DSS

Still another intersystem spectrum sharing has been studied within 5G. This is known as dynamic spectrum sharing (DSS), and it was introduced in Release 15 (2019) and enhanced after that. It allows an operator to put its 4G and 5G users into the same frequency band and thus enhance spectrum usage especially in the migration path from 4G to 5G. However, this is more related to the intraoperator scheduling of resource slots than actual DSA. Obviously, temporal dynamism of DSS is much higher than that of the other discussed DSA systems.

3.4 6G Spectrum Management Opportunities

Spectrum management within 6G could continue old ways. On the other hand, it could go in a more automated and flexible direction. The remaining of this chapter considers this new way. This section first discusses raised 6G use cases and then novel 6G techniques affecting spectrum management. The remaining sections consider frequency management for 6G.

3.4.1 6G Use Cases

It is important to briefly understand 6G use cases before discussing spectrum management. These are still wishes and aims presented in various white papers as well as in articles but also discussed in international organizations like the International Telecommunication Union, see Ref. [3]. These use cases are listed in Table 3.2. The numerical goals are at least 10-fold compared with 5G.

A goal derived from the use cases is to offer enhanced data rates compared with 5G, while another one is to offer improved service potential to verticals. In this sense, 6G will do for 5G the same what 5G did for 4G in areas that need higher throughput, i.e., boosting capacity and potential. Herein, smarter and more automated ways to manage spectrum in multi-radio access technology (multi-RAT) environments become important. Other goals discussed here arise from the sustainability target. First, 6G must be more energy efficient. This requirement

Table 3.2 The 6G use cases based on International Telecommunication Union (ITU) including comparison to 5G.

Enhanced from 5G	eMBB
	mMTC
	URLLC
New in 6G	Ubiquitous connectivity
	Integrated AI and communications
	Integrated sensing and communications
In high level	Ubiquitous intelligence
	Security/privacy/resiliency
	Sustainability
	Connecting the unconnected

eMBB denotes enhanced mobile broad band, mMTC massive machine type communications, ultra-reliable low latency communications (URLLC) ultrareliable and low-latency communications, and AI artificial intelligence.

is highlighted by predictions that data transfer is expected to consume more and more of produced energy if transported data grows as it does now. Certainly, this requirement will affect spectrum management algorithms within 6G as one of the key drivers.

Second, 6G should provide solutions to cut down the digital divide and connecting the unconnected and underconnected, i.e., societal sustainability. Reasons for this still occurring and challenges related to these problems especially in rural and remote areas are discussed in [4–6] and some references therein. A conclusion is that 6G should provide solutions that offer affordable (price) and sufficient (data rate and availability) services and, consequently, the system's price tag, and installation and maintenance costs should be kept low such that operators' willingness to invest in these areas would increase, or made other alternatives possible. This certainly affects the (to be) developed spectrum management algorithms. From the spectrum management point of view, rural and remote areas offer interesting possibilities since spectrum is often either unused or underused.

5G includes the nonterrestrial (NTN) part, but as a late comer, not optimized. This NTN trend is expected to be continued as an inherent part of 6G and, therefore, in an optimized fashion. Indeed, ubiquitous connectivity calls for this. The NTN part includes a satellite segment, high altitude platforms (HAPS), and occasionally unmanned aerial vehicles (UAVs or drones) to carry network elements. This means that spectrum will be managed in 3D, two horizontal dimensions and the vertical one, instead of traditional 2D way. This will add complexity to the spectrum management, especially if the NTN systems shared the common frequency band with the terrestrial elements of 6G. Furthermore, it is not only user equipments (UEs) that are connected, but more and more sensors and other Internet-of-Things (IoT) devices. In addition, backhaul solutions for remote access sites are needed. Traditionally, NTN (read satellite) systems have had, and still have, their own spectrum. If this trend was continued, 6G NTN spectrum management would be a different spectrum management entity continuing satellite communication (SATCOM) usage practices. However, if some NTN elements aimed to use the terrestrial spectrum, that would add complexity to spectrum management processes since coverage areas from even UAV altitudes would be much longer than what the most terrestrial base stations have. HAPS and satellites would have highly directional antennas but still coverage areas would be rather large on the surface of the Earth. Indeed, SATCOM systems have different tactics to organize beams at the Earth's surface: Earth-fixed beam (usually geostationary orbit [GSO]), quasi-Earth-fixed beam (usually GSO with steerable beams), Earth-moving (usually non-geostationary orbit [NGSO] without steerable beams), see, e.g., the NTN part of [7].

As a minor user group but still highly important, mission-critical users would benefit from having deployable (nomadic and operation on the static state) access points that can be set up fast in the case of emergency, if usual networks were not available. The trend of using civilian infrastructure by public authorities is a currently ongoing process as mission-critical users are moving to use 4G as their primary connectivity system in many countries during the 2020s. Mobile base stations would be even better since that would add mobility needed in some scenarios. Furthermore, small-scale user-to-user networks independent of base stations (i.e., ad hoc networks) would also be very beneficial for mission-critical users. Flexible, automated spectrum management would be a valuable feature in these systems. Today's systems include prioritization (i.e., reserving a part of allocated spectrum for mission-critical users if they urge it), but obviously, that sacrifices citizens connectivity although they might be eager to get more information about the situation. Consequently, flexible use of unused spectrum could be better primary solution and prioritization just a second one. Currently (as of October 2023), standalone backpack-style 5G base stations (including core) are available, such that this trend is growing.

3.4.2 6G Novelties

6G includes also technologies offering new possibilities and potential as shown in Table 3.2. These are integrated sensing and communications (ISAC) often accompanied by reflective intelligent surfaces (RIS) and, finally, the extensive use of artificial intelligence (AI) in the communications systems.

3.4.2.1 ISAC

As was detailed in Section 3.3.2, some spectrum sharing use cases call for a sensing system to support sharing. 6G is predicted to include ISAC functions where the sensing part could be used also for spectrum sensing in addition to other sensing tasks such as radar or radio-frequency-based object detection and recognition. This spectrum sensing part of ISAC could be a part of the sensing system used in spectrum sharing by forming improved spectrum situational awareness and allows cost-effective implementation of other spectrum sharing means than just CBRS.

As an example, since base stations and user devices will participate in spectrum sensing, the spectrum sensing situational picture would be more precise than what a sole (possibly more capable) sensor covering a large geographical area could provide. This would allow spectrum sharing in smaller geographical areas.

3.4.2.2 RIS

Reflective intelligent surfaces (RIS, just one of the used terms) means an array of elements that reflect radio waves in such a way that the reflection angle and other signal properties can be adjusted by adjusting elements' properties. This allow to point the reflected signal to areas that are beyond an object if seen from the base station. Naturally, this would complicate frequency management since the signal would propagate into an otherwise unreachable area. Spectrum management with RIS is considered in more detail elsewhere in this book.

3.4.2.3 AI

The attractiveness of AI for spectrum management lies in its ability to analyze large data masses and, therefore, get better understanding of local spectrum usage and channel models, which can be used in frequency allocation and resource usage optimization. However, due to its dependency on training data, AI methods are not necessarily suitable for dynamic spectrum management (DSM) situations where environment changes after training. We must better understand where AI methods are useful and where they are not.

3.4.3 6G Spectrum

Before diving into 6G spectrum management trends, it is good to look at 6G spectrum discussions. Spectrum is not directly related to spectrum management since the management should work in all used spectrum bands. However, it affects coverage area and used antenna technology (multiple input multiple output [MIMO] versus single sector antenna) that both influence the complexity of the spectrum management. Furthermore, the more available spectrum bands, the more possibilities for frequency management and usage optimization.

The discussion is herein based on Nokia's views explained in [8], but these reflect generic understanding. At the end, 6G will be used in many, if not all, frequency bands allocated for cellular communications. However, it needs an initial starting frequency (as 5G required 3.5 GHz band) and this was thought to be found between 7 and 15 GHz, which still offers reasonable coverage in addition to throughput. Another proposed new 6G band is around 100 GHz, i.e., sub-THz band, but this band is better suited for indoor usage.

3.5 Way Ahead Toward 6G Spectrum Management

It is foreseen that existing spectrum management means will continue to be used since they have been shown to be practical. However, system elements become more adjustable and capable. For example, new antenna arrays (MIMO technology) are more and more based on active antenna technology, meaning that a beam can be directed (also elevation in a limited manner) by adjusting its element phases. This opens room for denser spectrum allocation together with other interference management functions. However, this is not anymore a person-manageable task, but calls for automation. Furthermore, the 6G use cases call for more autonomy and dynamism from the spectrum management since spectrum usage must be accomplished in more complex, dynamic environment taking also other issues like energy saving into account. Mission-critical users would like to see automatic site set-up and adaptation if spectrum situation changes. Local network operators, who will provide the last mile service in rural and remote areas in many cases, do not necessarily have experts in their staff like traditional operators have. Consequently, automated plug-and-play type access points would be highly valued.

It can be concluded that 6G would benefit from a DSM system that provides automated and adaptable spectrum control that can interact with other local systems for handling interference problems. To this end, a brief literature survey is provided. It is followed by a high-level DSM architecture description for 6G. Regulators' role is important since they make DSM possible. Discussions about spectrum regulation can be found from chapter 2 of this book of this book. It is just good to know that several alternatives have been raised for sharing, including trading and leasing, as summarized in [9].

3.5.1 What Others Have Said?

The United Kingdom (UK) regulator OFCOM presented its view about dynamic spectrum access in the United Kingdom [10]. The report concluded that licensing process needs more automation and that better usage data would improve sharing processes. Furthermore, dynamic access best suits to boost capacity, not for primary access. Herein, dynamic access clearly does not mean automated processes to set the used frequency as meant in this chapter but the usage of spectrum opportunities. The report further concludes that also devices need to be more flexible and interfaces between databases and devices be open, i.e., DSM friendly. Overall, it was seen that DSM has the potential to play an important role in future cellular systems.

DSM and DSA algorithms have been studied for a long time now. The purpose herein is not to dig into these, since the aim is to stay at higher levels. However, these studies could serve as good starting points when particular DSM algorithms are developed for 6G. Keywords "6G" and "dynamic spectrum management" provided a few recent hits. Overview of DSA and blockchain's role in it is given in [11]. AI techniques for DSM have been studied in [12], where federated learning within NTN is considered, and in [13], where blockchain technology for IoT is analyzed. DSM, when base stations are at UAVs, is considered in [14]. There were also others, but the trend seems to be on using AI schemes for DSM (or DSA).

3.5.2 Generic DSM Architecture for 6G

The vision is such that an access point is a plug-and-play type object able to select used frequency in collaboration with other systems in the area and able to automatically change its frequency if spectrum situation changes (e.g., due to mobility or if new radio systems become in the area). The system should also be able to live with less capable (from the spectrum management point of view) systems.

Automated operation means that possible frequencies must be known. Furthermore, sharing possibility means that sharers must be known as well as sharing rules (free, negotiable, and fee). Finally, collaboration means that there must be a negotiation channel with other radios, networks, etc.

It is of upmost important that the overall system is kept as simple as possible such that it can be easily set up, managed, and maintained. Clearly, some information is very static but some may vary quickly. However, this depends on the use case. As an example, an access point in a rural or remote area results in a rather static situation since there are no other systems using these frequencies. A more dynamic use case is when a UAV-based access point, e.g., by public safety actors, will arrive in the area that readily has other radio systems.

Such a simple system readily exists since WiFi includes a rather simple DSM mechanism. Consequently, it provides a good initial example of what is meant by the DSM system (herein). WiFi radios know their possible frequencies (the two industrial, scientific, and medical [ISM] bands) and how to behave there (CSMA). They automatically select a carrier frequency and even tune it based on other WiFi systems. If a WiFi system (e.g., at home) contains several access points, they do not overlap in frequency. The first property is what is called a policy element in what follows. The second and last one is an intrasystem DSM element. Frequency tuning based on detected activity could be seen as a simple intersystem DSM element that uses spectrum sensing to avoid interference. A sharing element is missing from the WiFi system since it readily knows all available frequencies. It is clear that cellular systems need a somewhat more advanced DSM system since they operate on multiple frequency bands, sharing may be included and so on.

The proposed DSM system architecture block diagram including all the discussed elements is shown in Figure 3.2. This was initially discussed in [15] in the context of rural and remote area spectrum management. There could be several DSM-capable systems (A, B, and C in the figure) that could interact via intersystem DSM element though some systems may be less capable to operate in spectrum (C). Furthermore, there could be other noncollaborative radio systems and wireless backhaul in the area using the same potential spectrum. The spectrum sensing element is merely a needed capacity rather than a separate element. Indeed, it could be integrated or included in the radio system or be a separate system. The duties of the architecture building blocks are discussed next.

The policy block contains allowed spectrum, spectrum right owners, information about possibility to use spectrum by others (e.g., "yes, ask the spectrum right owner"), but also usage rules if such exist like in the ISM bands. The spectrum allocation information obviously come from the regulator and can also be

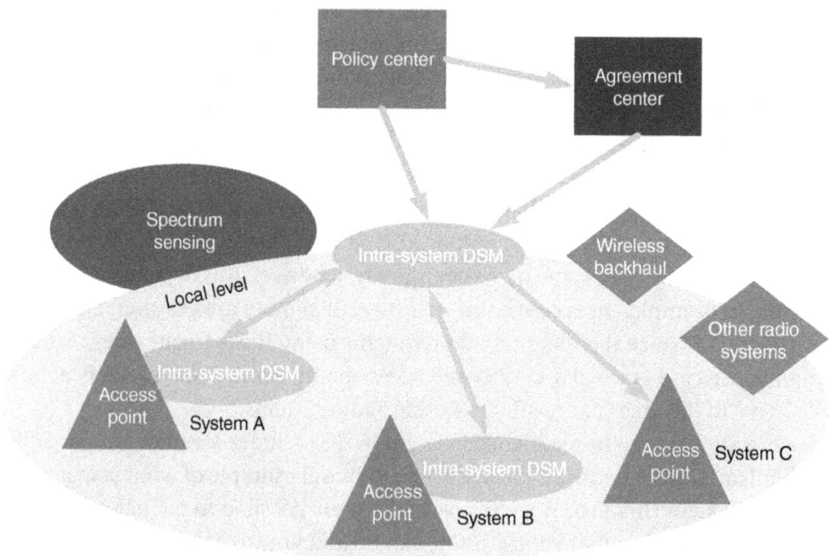

Figure 3.2 The block diagram of the proposed DSM architecture.

geographically limited. The block can also include optimization and usage criteria and restrictions that may come from the regulator but mostly from the operator itself. Consequently, this block is divided into subblocks that are set up and managed by different entities. The policy block also informs which frequencies are allocated for public safety operations if they would need additional capacity.

The sharing block includes the sharing rules (free, negotiable, and fee) of shareable frequencies and their geographical areas, though the latter may make the system too complex. The aim is that those who need spectrum can do negotiations through this automated entity. As an example, spectrum may be given free if a regulation demands it; it could be free based on exchange for spectrum in some other location, or there could be a fee for the spectrum since in many cases spectrum rights owners have paid large sums to have access to spectrum. There are several other possibilities too. Obviously, the systems (or their operators, indeed) get their spectrum from these two above blocks. The remaining blocks consider what to do with obtained spectrum.

The inter-system block deals with how to share available spectrum among systems. This could be from the point of view of an operator that has several systems in an area, but also interoperator collaboration could be possible though usually operators are orthogonal in spectrum. One potential use case would be interference management if spectrum was shared but spatial separation that has failed for some reason. Obviously, the systems must have a feedback channel for this purpose. Another use case occurs when secondary users are using the spectrum (e.g., CBRS tier three users), and it would be better to coordinate spectrum sharing in the given band than do avoid-interference-by-trial type sharing.

The intra-system block considers how to share allocated spectrum (for a particular system) among sectors and access points. This is traditional spectrum allocation, indeed. However, this should be highly automated in the future. Furthermore, it should take the set requirements into account, which could be time varying depending on the time of the day and location. For example, in an office area during work hours almost all resources might be needed, but outside office hours minimum service would suffice and others could be set to sleep. In remote areas, the desired coverage area could be just a few hundred meters (e.g., a small/dense village) or a few tens of kilometers (e.g., sparsely populated houses along a road and its side roads). Obviously, those spots could be best served using different frequency, transmit power, and so on. Numerous other needs can be imagined.

The system-level blocks could be one entity since their tasks are very dependent, especially if one operator is considered. If it was a local operator using shared frequency, then just the intrasystem level is needed though the intersystem level is needed to report interference issues to the spectrum owner.

The presented architecture is flexible. However, the details of its building blocks are pretty much unclear and need research and development actions before such a DSM system will be usable. Both flexible devices and (open) standardized interfaces are needed, as well as needed common database structures and their management tools.

It is obvious that in many nations current regulation does not allow taking the full benefits of the proposed architecture. Therefore, it is foreseen that first systems have very limited functionalities, but once the DSM system is demonstrated to be beneficial, new more flexible regulation may become reality. This should include changes to current frequency allocation, especially if "voluntary" sharing of allocated unused/underused spectrum does not work, since this spectrum is often needed in rural and remote areas for the last mile connectivity to bridge the digital divide.

3.6 Conclusion

Spectrum assignment for access points has been rather static. However, steps toward more DSM have been taken when CBRS and LSA were included. It was argued that 6G would significantly benefit from DSM or, indeed, 6G calls for DSM. More and more access systems will be available; they will be more flexible in spectrum and mobility of access points might be reality (e.g., UAV-based access points in some scenarios). Furthermore, 6G goals like connect the unconnected would benefit from automated installation, simple management, and maintenance. Automated spectrum management through DSM is a part of this solution. Finally, to support these ideas, a generic architecture for 6G DSM was introduced.

References

1 ETSI TS 103 235 V1.1.1 (2015-10) (2015). *Reconfigurable Radio Systems (RRS); System architecture and high level procedures for operation of Licensed Shared Access (LSA) in the 2 300 MHz - 2 400 MHz band*. European Telecommunications Standards Institute.

2 ETSI TS 103 652-2 V1.1.1 (2020-01) (2020). *Reconfigurable Radio Systems (RRS); evolved Licensed Shared Access (eLSA); Part 2: System architecture and high-level procedures*. European Telecommunications Standards Institute.

3 ITU (2023). ITU: 6G Standardisation Ready 'No Later Than 2030'. 6G World, June 29, 2023. https://www.6gworld.com/exclusives/itu-6g-standardisation-ready-no-later-than-2030/ (visited October 2023).

4 Saarnisaari, H., Dixit, S., Alouini, M.-S. et al. (eds.) (2020). 6G White Paper on Connectivity for Remote Areas. *6G Research Visions, No. 5*. University of Oulu.

5 Chaoub, A., Giordani, M., Lall, B. et al. (2021). 6G for bridging the digital divide: wireless connectivity to remote areas. *IEEE Wireless Communications* 29 (1): 160–168. https://doi.org/10.1109/MWC.001.2100137.

6 Saarnisaari, H., Chaoub, A., Heikkilä, M. et al. (2021). Wireless terrestrial backhaul for 6G remote access: challenges and low power solutions. *Frontiers in Communications and Networks* 2: 710781. https://doi.org/10.3389/frcmn.2021.710781.

7 3GPP TS 38.300 V17.6.0 (2023-09) (2023). 3rd Generation Partnership Project; Technical Specification Group Radio Access Network; NR; NR and NG-RAN Overall Description; Stage 2 (Release 17).

8 NOKIA (2023). Spectrum for 6G explained. https://www.nokia.com/about-us/newsroom/articles/spectrum-for-6G-explained/ (visited October 2023).

9 Digital Regulation Platform (2023). Spectrum management: key applications and regulatory considerations driving the future use of spectrum. https://digitalregulation.org/spectrum-management-key-applications-and-regulatory-considerations-driving-the-future-use-of-spectrum/ (visited November 2023).

10 OFCOM (2023). Opportunities for dynamic or adaptive approaches to managing spectrum in the UK: a discussion paper. https://www.ofcom.org.uk/__data/assets/pdf_file/0032/255956/discussion-paper-flexible-adaptive-spectrum.pdf (accessed 30 September 2024).

11 Muntaha, S.T., Lazaridis, P.I., Hafeez, M. et al. (2023). Blockchain for dynamic spectrum access and network slicing: a review. *IEEE Access* 11: 17922–17944. https://doi.org/10.1109/ACCESS.2023.3243985.

12 Lin, P., Meng, X., Liu, L., and Kadoch, M. (2022). Federated learning based 6G NTN dynamic spectrum access. In: *Signal and Information Processing, Networking and Computers* (ed. S. Sun, T. Hong, P. Yu, and J. Zou), 243–250. Singapore: Springer Nature Singapore. ISBN 978-981-19-4775-9.

13 Sun, Z., Liang, W., Qi, F. et al. (2021). Blockchain-based dynamic spectrum sharing for 6G UIoT networks. *IEEE Network* 35 (5): 143–149. https://doi.org/10.1109/MNET.101.2100138.

14 Wang, L.-C., Peng, H., Cheng-Sen Huang, A.C.-S., and Tsai, A.-H. (2021). Latest advances in spectrum management for 6G communications. *2021 30th Wireless and Optical Communications Conference (WOCC)*, 295–297. https://doi.org/10.1109/WOCC53213.2021.9603218.

15 Saarnisaari, H. (2023).6G local spectrum management for remote sites. *2023 IEEE Future Networks World Forum (FNWF).*

4

Artificial Intelligence-Enabled Dynamic Spectrum Management

Qiyang Zhao[1], Hang Zou[1], Yu Tian[1], Lina Bariah[1], Belkacem Mouhouche[1], Faouzi Bader[1], Ebtesam Almazrouei[2], and Merouane Debbah[3]

[1]*Digital Science and Telecom Research Centre, Technology Innovation Institute, Abu Dhabi, United Arab Emirates*
[2]*Artificial Intelligence Research Centre, Technology Innovation Institute, Abu Dhabi, United Arab Emirates*
[3]*KU 6G Research Center, Khalifa University of Science and Technology, Abu Dhabi, Shakhbout Bin Sultan St, United Arab Emirates*

4.1 Introduction

In this section we introduce the concept of dynamic spectrum allocation (DSA) for 6G and how machine learning will contribute to it. Being a scarce resource, spectrum is among the few elements in wireless communication that cannot be duplicated or enhanced. Wireless communication standards have continuously increased the efficiency of coding and modulation, increased the number of antennas using multiple input multiple output (MIMO) antennas. But for the spectrum, the only degree of freedom is to use higher frequencies toward millimeter waves and terahertz (THz) because the bit per hertz efficiency has become very close to theoretical limits. These high frequencies pose their own challenges to communication standards. The THz band has a unique molecular absorption loss, which is frequency dependent. Furthermore, the very short wavelength of THz signals makes it difficult for the signal to propagate for long distances since small objects can block it. As the industry is moving toward 6G, dynamic allocation of spectrum is vital to allow all the users to be served with the highest quality all the time as it is expected to be guaranteed by the future generation.

In the context of 6G, DSA is expected to use all the degrees of freedom in order to address the complexity of network demands. With intelligent networks, capable of handling different use cases and quality of services (Internet of Things (IoT), enhanced Mobile Broadband (eMBB), augmented reality/virtual reality (AR/VR)), a versatile spectrum approach becomes necessary. DSA's capacity

Intelligent Spectrum Management: Towards 6G, First Edition.
Edited by Sridhar Iyer, Anshuman Kalla, Onel Alcaraz López, and Chamitha De Alwis.

to accommodate varied traffic loads, application priorities, and geographical dynamics becomes highly important. However, this concept faces several challenges. The propagation characteristics differences between frequency bands, the interference created that can be difficult to predict, as well as the quality of service requirements that cannot be met are among the many considerations to take into account.

DSA for 6G can be seen as a generalization of frequency allocation in prior generations. The scheduler of a certain base station takes into account all the quality indicators and quality of service requirements from user equipment and dynamically allocates resources. This can be improved with link adaptation. In 6G, we might see a paradigm shift allowing the whole spectrum to be re-allocated in chunks between different operators and this allocation will take into account all the data traffic, the quality requirements, the environment, etc. The number of parameters to be managed grows significantly and cannot be handled using deterministic algorithms. This is where machine learning can be useful. Machine learning, a field of Artificial Intelligence (AI) enables systems to learn from a huge amount of data. Data-driven techniques can extract patterns within the spectral ecosystem. This tool can enable networks to predict demand fluctuations, proactively allocate frequencies in order to optimize usage. The combination of machine learning with DSA in 6G represents a progress toward self-built networks, sparing human intervention and enabling networks to continuously adapt to changing communication requirements.

In the sequel, we will discuss DSA and how it differs from static allocation. We will then give a brief overview of the widely used machine learning techniques and how they relate to spectrum allocation. Large language models (LLMs) and foundation models will also be discussed in relation to the dynamic allocation of the spectrum based on a large amount of data. Finally, we discuss the challenges ahead of the use of spectrum allocation for 6G and conclude the chapter with a summary of the main findings.

4.2 Dynamic Spectrum Allocation

Spectrum can be divided into two categories: licensed spectrum and unlicensed spectrum. Licensed spectrum is the scarcest resource needed by an operator or network manager and costs millions to billions of dollars. Licensed spectrum refers to frequency bands that are allocated by regulatory authorities (such as the Federal Communications Commission (FCC) in the United States, OFCOM in the United Kingdom, Agence Nationale des Fréquences (ANFR) in France) to an operator usually through an auction. Once the band is licensed, the operator has the exclusive rights to use these specific frequency bands for their customer, or license them to third parties either temporarily or permanently. To highlight this complexity, we show the figure of the frequencies allocated by the FCC for different uses below. For the sake of readability, in Figure 4.1, we limit the spectrum from 40 to 74 GHz. The overall chart can be found on the FCC website.

Having a licensed spectrum is very expensive, but brings an unmatched level of reliability since the operator fully controls the band. So the likelihood of interference is almost nil (except from malicious or unauthorized interferer). This exclusivity also leads to a high quality of service and reliable performance since the quality of service estimation does not suffer from any unknown interference contribution.

Unlicensed spectrum, on the other hand, refers to frequency bands that can be used by anyone without the need for a prior license. These bands are usually used for short-range communication such as WiFi and bluetooth. Another use is the private networks that can locally use the band especially in isolated areas without affecting other users. Unlicensed spectrum is free, and thus can be used by smaller entities and individuals for innovation and testing new ideas without going through the lengthy process of requesting a license.

Unlicensed spectrum also provides flexibility as the absence of the need to pass through the licensing hurdle enables easy and quick device deployment. Cost-efficiency is another positive aspect of unlicensed spectrum since there are no upfront costs to be borne before starting to use the required bands.

For licensed spectrum, the frequencies allocated in a fixed way take a long time to be assigned to the corresponding operators. Furthermore, it was shown that the spectrum is underutilized and that many parts of the spectrum remain unused (usually called white spaces) [1]. For example, it was shown that in the band from 30 to 2900 MHz, 82.6% of the allocated spectrum is unused. This means that the problem is the way spectrum is accessed and not the absolute shortage of the spectrum. The fact that some spectrum is reserved does not contribute to the efficient use of resources. With the ever increasing demand for higher data rates in current and future cellular systems, a new paradigm in the spectrum allocation is needed. This paradigm is called DSA [2]. The dynamic allocation can solve many problems in time and space where the same spectrum is not needed all the time or everywhere, whereas the fixed allocation cannot. An operator can decide to automatically allocate spectrum to the radio access technology that needs it the most (e.g., 4G system can create space to a 5G system when a higher demand is observed). In this way, a smooth transition is guaranteed where the coexistence of the two systems leads to a continuous service until the new system completely takes over. Another form of dynamic spectrum is the dynamic allocation between operators. This is a more complicated issue because of the commercial and technical difficulties that make it more theoretical than practical. This issue will be discussed later in this chapter.

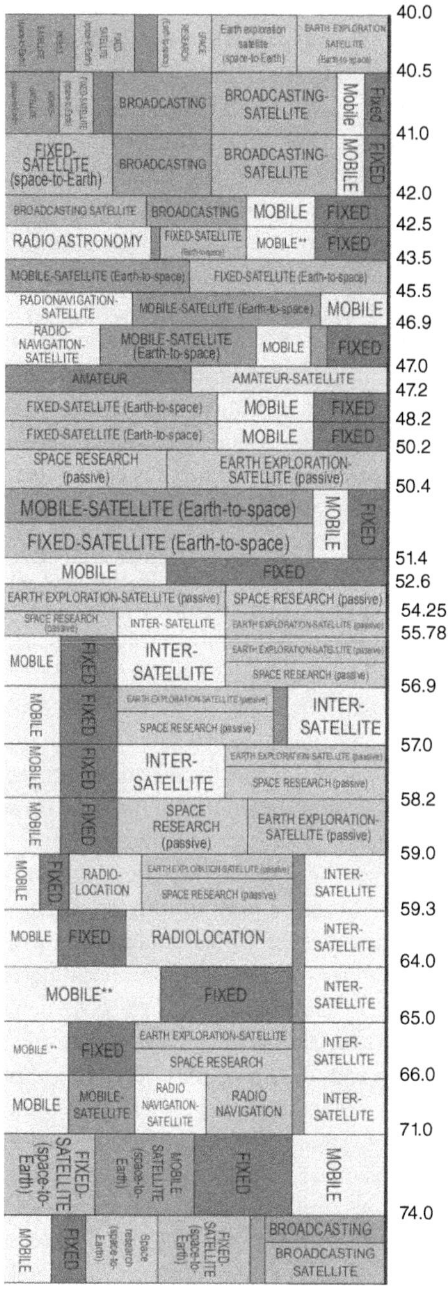

Figure 4.1 The United States radio-frequency allocation chart between 40 and 74 GHz.

Dynamic spectrum access and dynamic spectrum allocation might be slightly different [3]. Dynamic spectrum allocation assumes the existence of a central entity allocating the spectrum to different nodes asking (sometimes bidding) for the spectrum. Dynamic spectrum access, on the other hand, presupposes that the spectrum is already allocated and that the nodes are opportunistically sensing the already allocated spectrum in order to fit their communication within. Dynamic spectrum access is usually associated with cognitive radio in which transceivers can detect whether a communication channel is used or not as well as the radio technology being used in this channel. In this chapter, we will use the two terms interchangeably.

As shown in Figure 4.2, there are mainly three models for DSA [3]:

1. **Dynamic exclusive use model**: In which spectrum bands are licensed for specific users and for an exclusive use, but flexibility is introduced to improve spectrum efficiency. Two sub-approaches exist under this model:
   - spectrum property rights, which allows licensees to sell and trade spectrum and to freely choose technology. Economy and market will thus play a more important role in driving toward the most profitable use of this limited resource.
   - dynamic spectrum allocation, which aims to improve spectrum efficiency through dynamic spectrum assignment by exploiting the spatial and temporal traffic statistics of different services. Similar to the current static spectrum allotment policy, such strategies allocate, at a given time and region, a portion of the spectrum to a radio access network for its exclusive use. This allocation, however, varies at a much faster scale than the current policy.
2. **Open sharing model**: Which employs open sharing among peer users as the basis for managing a spectral region. This is compared to what is used in unlicensed bands with technologies WiFi.
3. **Hierarchical access model**: Which is built upon a hybrid use of the above two methods in which we have primary and secondary users. The basic idea is to open licensed spectrum to secondary users at the condition that the interference perceived by primary users (licensees) will be limited. Two approaches to spectrum sharing between primary and secondary users have been considered:
   - **Spectrum Overlay**: The spectrum overlay model targets aggressive opportunistic exploitation of white-space or spectrum gaps in spatial-temporal domain. The secondary users will have to identify the idle spectrum band, which is not used by licensed users at a certain time and location and use those idle spectrum bands dynamically to unlicensed users. Spectrum overlay example include Time-Division Multiple Access (TDMA),

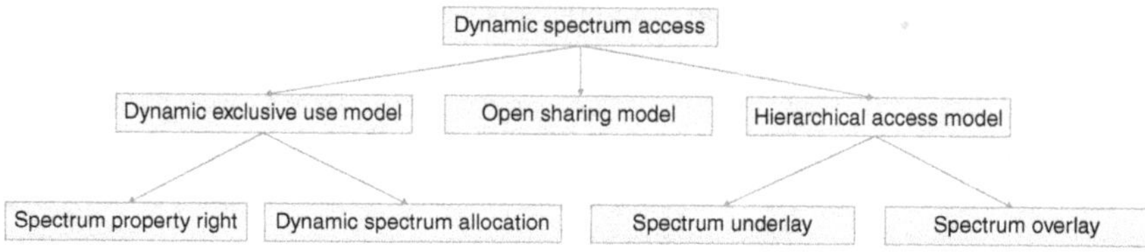

Figure 4.2 Dynamic spectrum access categories.

Frequency-Division Multiple Access (FDMA), and Orthogonal Frequency-Division Multiple Access (OFDMA) system.

- **Spectrum Underlay**: In this mode, some constraints are imposed on the maximum transmission power of secondary users in order to keep the overall interference below the noise floor of primary users. The spread of transmitted signals over an ultra wide band (UWB) frequency enables secondary users to achieve a short-range high data rate with very low transmission power.

DSA is implemented in many standards such as Institute of Electrical and Electronics Engineers (IEEE) IEEE 802.22, IEEE P1900.4, and the Third Generation Partnership Project (3GPP). For example, in the case of 3GPP, DSA enables a smooth transition from long-term evolution (LTE) to new radio (NR). It also enables new 5G services to be deployed and to coexist with LTE [4]. In addition to that, effective use of spectrum is guaranteed. DSA enables a single Legacy-LTE spectrum to be used as a carrier for both LTE and NR services . The most forward way to migrate from LTE to NR is to use spectrum refarming, i.e., the free parts of the spectrum are gradually being directed and used by the new standard as shown in Figure 4.3.

DSA is a better option as it enables the two technologies to exist together and balance the load of each one during the transition period. When the NR becomes a clear dominant then a complete spectrum shift can be performed as show in Figure 4.4.

In addition, the use of millimeter wave for 5G causes the coverage to shrink because of the millimeter wave propagation properties. This issue can be alleviated by implementing DSA on low-band carriers and using carrier aggregation in order to use the low-band carrier with the higher-band carrier.

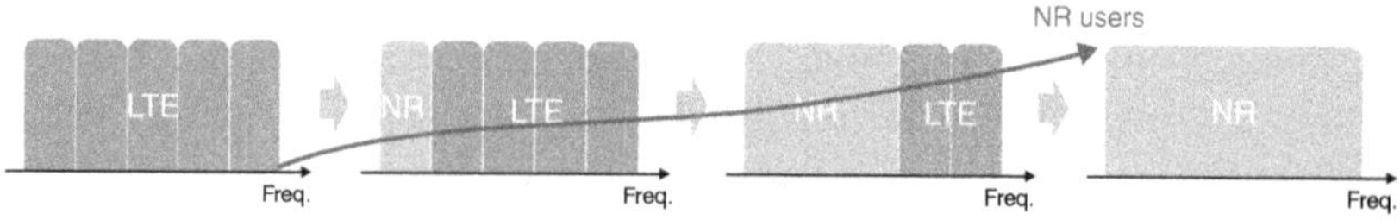

Figure 4.3 Spectrum refarming as a smooth transition method from LTE to NR. Source: Samsung [4]/Samsung Electronics Co., Ltd.

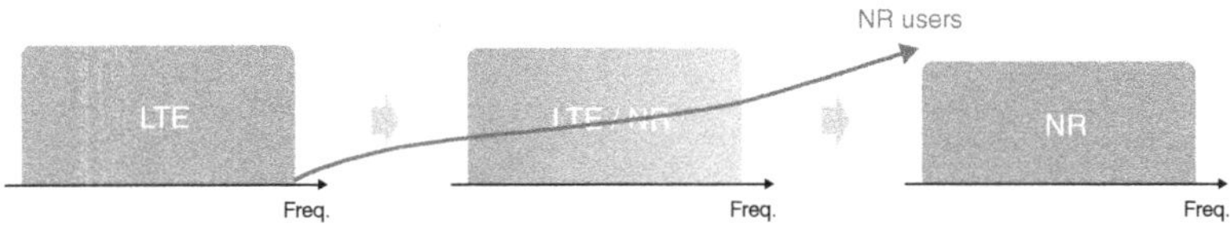

Figure 4.4 Dynamic spectrum allocation as a more efficient smooth transition from LTE to NR. Source: Samsung [4]/Samsung Electronics Co., Ltd.

4.3 Machine Learning for Dynamic Spectrum Allocation

The use of AI in dynamic spectrum management has been widely studied since cognitive radio [2], which permits the secondary users to utilize the blank spectrum resources in time, frequency, space, power from the primary user's spectrum band, such as the TV White Space band in IEEE 802.22 Wireless Regional Area Networking standard. In doing so, interference or collision could occur when the users transmit on the same spectrum resources. Spectrum sensing plays a key role in cognitive radio, such as measuring the received power at different spectrum band. However, this faces severe challenges in terms of spectrum sensor device complexity, availability, and energy consumption. In this context, AI has been widely used to empower smarter spectrum utilization, as shown in Figure 4.5. On one hand, it can predict the spectrum usage at different time, frequency, space based on the past transmission pattern, including the user traffic and mobility. On the other hand, it can make effective decision in selecting appropriate resources, time slot, and transmit power that can maximize the fairness or an expected performance between different users.

A number of AI mechanisms has been developed for spectrum management [5]. Multiarmed bandit (MAB) has been extensively studied for radio resource allocation [6]. In such algorithm, multiple users allocate a fixed limited set of radio resources in a competing radio environment, to maximize their expected gain, such as throughput, latency, or reliability. The global resource allocation strategy is partially known to each specific user, through sensing its surrounding radio environment. The algorithm improves the allocation strategy toward the expected gain as the transmission iteration passes. MAB is a classic reinforcement learning (RL) problem that exemplifies the exploration and exploitation trade-off dilemma. For example, the users could start from greedy allocation at the beginning and exploit more past experiences as with more iterations, and finally converge to an optimized strategy where the spectrum is well partitioned between different users, so as to minimize interference or collisions. Contextual bandit is an extension to MAB, which exploits the environmental states, such as the transmit power, data load, and user location [7]. It can explore multiple optimal strategies in different

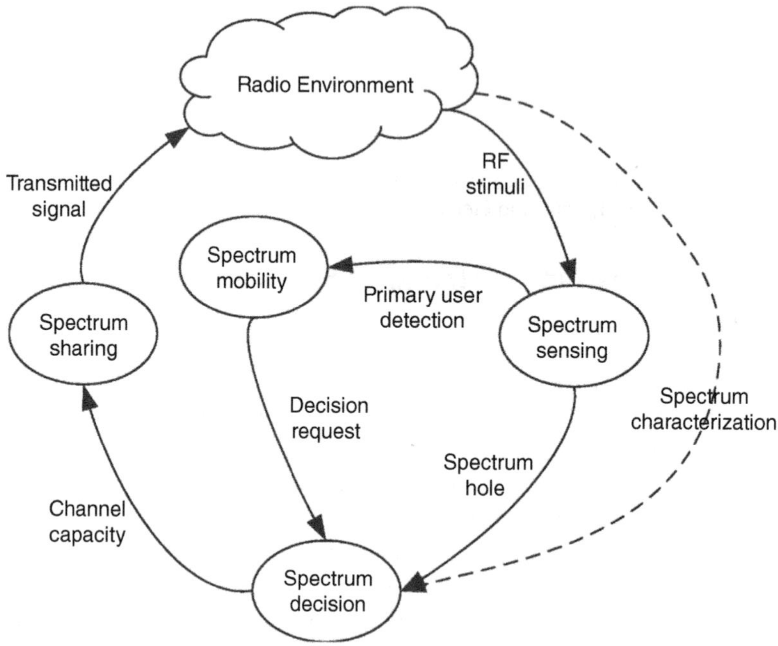

Figure 4.5 Cognitive radio cycle. Source: Adapted from Mitola and Maguire [2].

scenarios, which practically performs better than classical MAB approaching a single strategy. Moreover, combinatorial bandit has been utilized for hierarchical setup, such as decision makers in one base station with several users [6]. It allows cooperation between multiple agents so as to maximize the network-level performance.

Whilst MAB and its variants are simple and effective in radio resource allocation for spectrum management, they lack modeling or exploring the state transitions, such as the reward of the action in the next state. This may lead to the agent falling into a local optimal, where the best action for the current state could be poor for the future states. For example, the users in a network selecting the best resources at current time slot may suffer from high interference in the next slot due to high load and interference caused to each other, making the network difficult to converge. RL has been widely adopted to spectrum management. Specifically, the agent's action selection is modeled as a map called policy π, which gives the probability of taking an action a in a state s. The value function defines expected return starting with the state s, and successfully following policy π. It estimates how good it is to be in a given state. The reward function evaluates the instantaneous performance of the action. The accumulated reward in the value function evaluates the long-term performance of taking an action in a state [8].

Classical RL algorithms including State-Action-Reward-State-Action (SARSA) or Q learning have been widely used in spectrum allocation [9]. The RL agent can be modeled as a base station, user, or network taking actions on selecting radio resources in time, frequency, space, and power domain. The reward function can be defined as an objective of throughput, delay, collisions, transmission errors, interference, etc. An RL agent interacts with the radio environment and updates the resource allocation policy. SARSA is an on-policy RL algorithm where policy update is based on the actions taken. On the other hand, Q learning is an off-policy RL algorithm, which estimates the optimal state-action value function based on the maximum reward of available actions [10]. RL is modeled as a Markov decision process (MDP), which is suitable for the spectrum management problem in wireless communication systems [9]. One can see state transition as the spectrum resources being taken by a user, leading to a new state of available resources or the changed interference environment. RL learns optimal actions at each state leading to a converged optimal state, which can be seen as the maximum throughput or minimum interference among network users, depending on the targeted goal. Such goal should be represented in the reward function of RL agent taking action at each state. Furthermore, exploration and exploitation trade-off policy has a key impact on the RL performance. The agent selects action based on a stochastic policy at exploration stage, and on the learnt experience (i.e., Q value) during exploitation stage. The exploration policy can be ϵ-greedy, Boltzmann, upper confidence bound, etc., which gives the RL agent a probability to select actions outside the learnt policy. It is crucial to guarantee convergence to an optimal policy especially in a dynamic environment.

A challenge of RL in complex scenario is the large state and Q table. Deep RL has been studied to overcome this issue [11]. In dynamic spectrum management, the state parameter of received radio power, traffic load, packet size, and connected users are continuous value. Traditional Q learning should quantize these parameters in order to build an MDP of multiple states. However, this could constrain the agent exploring intermediate states, and result in a large Q table. Deep learning has shown effective clustering of states in RL by training the neural network with the reward as labels. Classical deep RL includes value based, policy gradient, and actor critic, with a number of variations in each. The value-based deep RL uses neural network to represent the Q table. It produces the Q value of each action at a given input of state. The stochastic gradient descent is applied to approximate the reward value given by the reward function. This is suitable for a discrete action space, such as assigning OFDM physical resource blocks (PRBs), which are based on predefined transmission time intervals (TTI) and subcarrier spacing (SC). On the other hand, the policy gradient method is suitable for a continuous action space. The expectation of the product of the reward and gradient of the log of the policy is used to update the neural network. This allows the neural network to directly output the value of the actions, which eliminates the need of discrete actions in value based method. The example applications in wireless network includes power allocation, load balancing, flow control, etc. Similar to the value-based method, policy gradient selects the maximum Q value at each iteration. However, an issue in both methods is that a small change of Q value could largely affect the actions being taken, which can cause poor performance when the agent has not gained sufficient experiences. The actor-critic method is proposed to utilize a critic network to evaluate the actions toward the global goal. The goal of actor is to maximize Q value while the goal of critic is to minimize the time-difference error of environment rewards and its predict Q value on the selected action.

The challenges of applying RL in spectrum management are as follows. First, RL relies on a well-defined reward function to model the impact of each state-action pair toward the goal. It usually lacks a theoretical proof of the optimal state and reward definition with multiple key performance indicators (KPIs). For example, with the final goal of maximizing throughput and minimizing delay in the network, arbitrary parameters should be placed on these KPIs in the reward function, which can be in different scales and leads to different results. Second, RL is a delayed reward process where the rewards are given only after the actions have been taken. For example, the throughput can be measured only after the packet has been transmitted. This can cause poor impact to the network, such as interference, congestion, and collisions. Third, RL requires a large amount of data collection through exploration—exploitation process. In each iteration, the actions taken by RL agent can cause impact to the performance. It requires large experience replay to avoid falling into local optimal. Finally, the ultrahigh reliability services target reliability at 99.9999%, which cannot afford the trial-and-error learning. The measurement of state and reward can also cause high energy consumption and complexity to the network devices, such as sensing the empty spectrum.

In order to overcome these challenges, the model-based RL learns an environment model and uses it to control the decision policy [12]. The model represents the environment dynamics, such as the state transition probabilities. During model-based policy optimization, the environment model plans a sequence of actions toward the targeted state. In each iteration, the first planned action is executed and the resulting state is used to minimize its difference to the planned state with a dynamic model, which optimizes the policy from the environment model. With this approach, RL can effectively utilize experience to improve sample efficiency and reduce data collection. However, the challenge with model-based RL is to learn via an environment model, which degrades its converged performance compared to model-free RL. In this context, the concept of world model is proposed, which is an expert model trained on large environmental data [13]. It can thus perform planning for various tasks to guide the agents. Such experiences can be seen as the common sense knowledge of radio spectrum allocation in a region, or basis of radio propagation and interference environment. This can largely accelerate model-free RL-based spectrum management, and give agent the capability to generalize the spectrum allocation policy in unseen scenarios.

4.4 Large Language Models for Dynamic Spectrum Allocation

LLMs are extremely large-scale Transformer-based neural networks trained on gigantic amounts of text data. Here we provide a brief-but-comprehensive overview of recent advances in LLMs.

Almost all modern LLMs are based on Transformers [14]. With the help of the multi-head attention, Transformer could capture long-range dependencies in parallel by having every word attending all previous words or every word except the target. A complete Transformer is composed of two parts, i.e., encoder and decoder. LLMs could utilize only the encoder/decoder part or both of them. Bidirectional encoder representations from transformers (BERT) [15] and its variants utilize merely the encoder part of Transformer. LLMs such as GPT-4 [16] using only the auto-regressive decoder part are becoming dominant and mainstream recently. LLMs applying both encoder and decoder part exists, e.g., T5 [17].

Compared to conventional deep-learning approaches, LLMs follow different paradigm of training, i.e., pretrain-then-adapt. LLMs are first pretrained on vast amount of **unlabeled** texts for given pre-training tasks such as causal language modeling or denoising autoencoding. Then LLMs are fine-tuned or adapted to downstream tasks through different methods, e.g., task dataset fine-tuning, instruction tuning, and alignment tuning. Besides, prompt engineering, i.e., carefully designing the form and contents of a prompt, serves as a practical adaptation technique when parameters of LLMs are too costly to update or unavailable. We list some key features of LLMs for DSA: (i) in-context learning [18]: LLMs are few-shot learners, i.e., LLMs are capable to learn a new task from a few prompted examples in the context; (ii) system 2-type reasoning: LLMs are shown to be able to perform tasks involving reasoning and logical deduction. Chain-of-thoughts (CoTs) [19] and its variants are reasoning techniques, which solve problems through a specific structure of process in multiple intermediate steps; (iii) instruction following: fine-tuning LLMs on a mixture of multi-task datasets are shown to be able to follow instructions in natural language without explicit examples [20]; (iv) multimodal learning: LLMs such as contrastive language-image pre-training (CLIP) [21] and AudioPaLM [22] can deal with data in multimodality form, e.g., text, image, and audio. Given above advantages, we believe that LLMs are extremely suitable and useful to enable DSA in upcoming 6G era. We envision on some potential use cases of LLMs on DSA.

- **Multimodality Spectrum Sensing and Allocation**: Data and information in different modalities exist in networks. Multimodality LLMs could process radio frequency (RF) signals, images taken from cameras and signal from other sensors to gather useful information to determine the spectral environment. Then the spectrum allocation task could be decomposed into parallel (sequential) sub-tasks modeled as common (optimization) problem with known solving techniques.
- **Network slicing and customization**: LLMs can support network slicing service, where different virtual networks are created on the same physical network tailored for specific use-cases, e.g., Internet of Things (IoT), AR/VR, and Vehicle-to-everything (V2X). LLMs can dynamically allocate spectrum for these use cases based on the specific requirements of each use case.
- **Multiagent connected intelligence**: Equipped with lightweight LLMs, agents can work collaboratively to render spectrum allocation intelligent. Short messages containing essential information of agents' state and action in natural language which are summarized by LLMs could be sent among relevant user equipments (UEs). Agents are able to decide and plan in a reasonable manner based on their perceived information and received messages with the assistance of LLMs through reasoning and planning. This multiagent system with connected intelligence could potentially avoid collisions in spectrum allocation and reduce the latency of the network.
- **Security and spectrum spoofing detection**: Spectrum spoofing refers to malicious activities where attackers mimic legitimate devices or network nodes to gain unauthorized access. Spoofing is usually difficult to be detected since the difference between normal and abnormal behaviors is minor. Pre-trained LLMs could be further fine-tuned on different datasets which contain both normal data and spoofing data identified historically. After fine-tuning, LLMs could recognize potential spoofing attempts.

Despite the potential of integrating LLMs in DSA, various challenges still exist:

- Low-latency Decision-Making: Inference time of current LLMs could be too long for the low-latency requirement of DSA for real-time constraint and vast amount of spectrum data.
- **On-device LLM**: Scale of current LLMs such as GPT-4 is huge to operate on a typical UE. Therefore, current model compression techniques should be adapted to LLMs without damaging their emergent abilities such as in-context learning.

- **Hallucination Management**: Hallucination refers to the behavior of LLMs producing nonsensical or untruthful contents w.r.t. provided source or common-sense knowledge. Hallucination of LLMs such as illegal use of licensed spectrum and reserved spectrum would be detrimental to existing networks.
- **Continuous Learning**: As the network evolves, e.g., regulation and policy changes, LLM-based DSA are expected to adapt accordingly. Regular knowledge injection is necessary for LLMs through efficient fine-tuning.
- **Energy Efficiency**: Pretraining or fine-tuning of LLMs is known to be power-hungry. For example, LLaMA in [23] as the most popular open-source LLM, consumed merely 449 MWh for training. How to reduce the power during training/inference/fine-tuning is essential to achieve energy-efficient DSA.
- **Transparency and Explainability**: Without a widely accepted explainability theory of deep neural network (DNN), it is even more challenging to understand how and why an allocation strategy is adopted by LLMs with extremely large scale. The black-box nature of LLMs obstructs the diagnosis and problem-solving in case of anomaly.
- **reliability**: Being essentially probabilistic models, the decision policy obtained by LLMs could be nondeterministic and unstable. In summary, there is a long path to overcome above difficulties before applying LLMs to DSA.

4.5 Challenges and Future Directions

Machine learning has been studied in dynamic spectrum management for decodes. However, the implementation in the real Telecom network is still limited. For example, 5G dynamic spectrum sharing is still limited to between 4G and 5G frequency-division duplexing (FDD) bands [24]. There are many challenges and open issues to be addressed toward a fully shared spectrum:

- **System Architecture**: Enabling dynamic spectrum management requires a redesign of existing telecom system cross multiple layers. The majority state-of-the-art research focuses on spectrum sensing and resource allocation. This requires considerable changes in the RF, physical, and MAC layers. For example, when performing received signal power measurement, a common control channel cross the spectrum band is needed. This will change the spectrum and frame structure design cross multiple systems. Moreover, scheduling between multiple bands requires complex signaling. 3GPP system relies on a interference coordination between multiple cells and bands. Dynamic spectrum management will change all the underlying protocols designed for resource allocation.
- **Reliability**: 5G system aims to deliver up to 99.9999% reliability in ultra-reliable low latency communications (URLLC) services, such as autonomous vehicles

or health cares. The majority of nowadays machine learning (ML) models are probabilistic model, which results in higher error probability than the URLLC requirement. This is even more obvious in the generative AI models (i.e., LLM), where the outputs are more uncertain. It could cause harmful interference in the network while AI making poor decisions. Moreover, the popular RL algorithm requires exploration and exploitation to converge, which also makes the system unstable.

- **Energy Efficiency**: The energy consumption of dynamic spectrum sharing includes the energy used for sense the spectrum, train the AI model and inference. Sensing requires base station (BS) or UE in active model, which consume a considerable amount of energy in the RF chain, especially measuring a wide spectrum band. 5G is designed to minimize the wake-up time for channel measurement to save energy, which can be more important than the gain from spectrum efficiency. Training a large AI model can consume ultra-high energy including both computation and data collection. Moreover, large AI models are highly complex to be deployed on energy constrained-network devices.
- **Latency**: The latency includes the time for sensing the spectrum and making decisions from inference. Spectrum usage is dynamically changing while interference can be detected reactively. This can cause interference to data transmission. The delay of large AI model in producing output can also increase the data delivery time, which is unsuitable for the ms level delay requirement in 5G.
- **Generalization**: The majority of deep learning algorithms are designed for independent identical distributed data, which performs well in a scenario with the same data distribution as the training dataset. However, this causes significant challenges to spectrum sharing where collecting spectrum data is highly expensive in terms of complexity and energy. The unbalanced distribution of data across different spectrum bands can cause poor generalization of AI models in unseen scenarios. For example, a model trained in rural area can hardly be used in the urban area.
- **Convergence**: The ML model is trained to approximate a function to partition the data. To guarantee convergence, the training dataset should be stable. Most of the online learning algorithm assumes the agents play repeatedly in a stable environment, so as to converge on an optimal policy. However, in wireless communications, it is challenging to emulate a stable radio environment and allow the mobile devices to perform a large amount of actions. The dynamics of the system make it difficult to explore an optimal policy with a few explorations. Moreover, the number of devices in wireless network is at million level, making it hard to converge on an optimal solution satisfying different goals.

4.6 Conclusion

In this chapter, we provided holistic reviews and perspectives on dynamic spectrum management with machine learning. We first presented our perspective of the role of dynamic spectrum access and artificial intelligence in 6G. Following this, we studied the spectrum allocation policies in different radio systems cross various regions. The benefits and costs of licensed and unlicensed spectrum in supporting radio communications were compared. We then introduced the need for dynamic spectrum allocation to improve the utilization of limited spectrum resources, such as coexistence of different generations of cellular systems. Different dynamic spectrum access models from dynamic exclusive use, open sharing to hierarchical access were analyzed. Furthermore, we summarized the current progress of dynamic spectrum sharing between 4G and 5G systems.

Subsequently, we introduced the research and implementation of machine learning in dynamic spectrum management. We started from the concept of cognitive radio where machine learning plays a key role in decision making on resource allocation from sensing the spectrum holes. Classical algorithms from multi-armed bandit to reinforcement learning were discussed, including their applications in selecting or assigning time, frequency, and power resources. Furthermore, we analyzed the issues of these algorithms in the communication systems, such as the cost of exploration and exploitation, data sampling, and convergence. This could have harmful impact to interference, energy consumption, throughput, and reliability.

To overcome these issues, we introduced the need for a world model encoding the background knowledge and experiences with capability of planning and reasoning, where the recent techniques of generative AI and LLM can play an important role. We first introduced different architecture of LLM from the use of transformer. Followed by this we described different training and fine-tuning approaches. We then emphasized the advantages of LLM compared to traditional AI models, including system-2-type reasoning, multitask, multimodality, and generalization. With multimodality LLM pretrained on RF signals, different downstream radio communication tasks in dynamic spectrum access system can be performed with a few-short prompts. The LLM-powered autonomous agent can automate the process of time, frequency, power allocation to trigger configuration and optimization of different network protocols and hardware, which can be achieved by fine-tuning or augmenting LLM with Telecom domain knowledge. In this context, LLM can break down higher-level goal (i.e., targeted QoS) into lower-level actionable tasks (i.e., tuning power amplifier, changing scheduling scheme), and interface with related functions. Moreover, LLM can deduce the logical relation between a goal and policy from the environment, which has the potential to generalize AI in dynamic spectrum management to diverse scenarios and supporting emerging service requirements.

The pathway to an AI native wireless communication system with dynamic spectrum management is still challenging. There is a large gap from theoretical performance in simulators to the practical use with severe requirements of reliability, energy efficiency, and implementation cost. Significant advancements in research and development of both AI and telecom systems are needed to realize a fully autonomous dynamic spectrum management system.

References

1 Akyildiz, F., Lee, W.Y., Vuran, V., and Mohanty, S. (2006). Next generation dynamic spectrum access cognitive radio wireless networks: a survey. *Computer Networks: The International Journal of Computer and Telecommunications Networking* 50 (13): 2127–2159.

2 Mitola, J. and Maguire, G. (1999). Cognitive radio: making software radios more personal. *IEEE Personal Communications* 6 (4): 13–18.

3 Zhao, Q. and Sadler, B.M. (2007). A survey of dynamic spectrum access. *IEEE Signal Processing Magazine* 24 (3): 79–89. May

4 Samsung (2021). *Technical White Paper Dynamic Spectrum Sharing.*

5 Grace, D. and Zhang, H. (eds.) (2012). *Cognitive Communications: Distributed Artificial Intelligence (DAI), Regulatory Policy and Economics, Implementation.* USA: Wiley-Blackwell.

6 Zuo, J. and Joe-Wong, C. (2021). Combinatorial multi-armed bandits for resource allocation. *2021 55th Annual Conference on Information Sciences and Systems (CISS)*, 1–4. IEEE.

7 Qureshi, M.A. and Tekin, C. (2020). Fast learning for dynamic resource allocation in AI-enabled radio networks. *IEEE Transactions on Cognitive Communications and Networking* 6 (1): 95–110.

8 Sutton, R. and Barto, A. (1998). *Reinforcement Learning: An Introduction, Adaptive Computation and Machine Learning Series.* MIT Press.

9 Zhou, F., Lu, G., Wen, M. et al. (2019). Dynamic spectrum management via machine learning: state of the art, taxonomy, challenges, and open research issues. *IEEE Network* 33 (4): 54–62.

10 Watkins, C. and Dayan, P. (1992). Q-learning. *Machine Learning* 8: 279–292.

11 Mnih, V., Kavukcuoglu, K., Silver, D. et al. (2013). Playing Atari with deep reinforcement learning. *arXiv preprint arXiv:1312.5602.*

12 Moerland, T.M., Broekens, J., Plaat, A., and Jonker, C.M. (2022). Model-based reinforcement learning: a survey. *Foundations and Trends® in Machine Learning* 16 (1): 1–118.

13 Ha, D. and Schmidhuber, J. (2018). World models. *arXiv preprint arXiv:1803.10122.*

14 Vaswani, A., Shazeer, N., Parmar, N. et al. (2017). Attention is all you need. In: *Advances in Neural Information Processing Systems*, 5998–6008.

15 Devlin, J., Chang, M., Lee, K., and Toutanova, K. (2019). BERT: Pre-training of deep bidirectional transformers for language understanding. *Proceedings of the 2019 Conference of the North American Chapter of the Association for Computational Linguistics: Human Language Technologies, NAACL-HLT 2019, Minneapolis, MN, USA (2-7 June 2019), Volume 1 (Long and Short Papers)*, 4171–4186. Association for Computational Linguistics.

16 OpenAI (2023). GPT-4 Technical Report, GPT-4 Technical Report.

17 Raffel, C., Shazeer, N., Roberts, A. et al. (2020). Exploring the limits of transfer learning with a unified text-to-text transformer. *Journal of Machine Learning Research* 21: 1401–14067.

18 Brown, T.B., Mann, B., Ryder, N. et al. (2020). Language models are few-shot learners. In: *Advances in Neural Information Processing Systems 33: Annual Conference on Neural Information Processing Systems 2020, NeurIPS 2020 (6–12 December 2020) Virtual.*

19 Wei, J., Wang, X., Schuurmans, D. et al. (2022). Chain-of-thought prompting elicits reasoning in large language models. In: *Advances in Neural Information Processing Systems 35*, 24824–24837.

20 Wei, J., Bosma, M., Zhao, V. et al. (2021). Finetuned language models are zero-shot learners. *International Conference on Learning Representations.*

21 Radford, A., Kim, J.W., Hallacy, C. et al. (2021). Learning transferable visual models from natural language supervision. *International Conference on Machine Learning*, 8748–8763. PMLR.

22 Rubenstein, P.K., Asawaroengchai, C., Nguyen, D.D. et al. (2023). AudioPaLM: A large language model that can speak and listen. *arXiv preprint arXiv:2306.12925.*

23 Touvron, H., Lavril, T., Izacard, G. et al. (2023). LLaMA: Open and efficient foundation language models. *arXiv preprint arXiv:2302.13971.*

24 Sharma, S.K., Bogale, T.E., Le, L.B. et al. (2018). Dynamic spectrum sharing in 5G wireless networks with full-duplex technology: recent advances and research challenges. *IEEE Communications Surveys & Tutorials* 20 (1): 674–707.

5

Infrastructure for Spectrum Management Enabled by Virtualization and Network Slicing

Uditha Wijewardhana, Nishan Dharmaweera, and Bhathiya Pilanawithana

Faculty of Engineering, University of Sri Jayewardenepura, Nugegoda, Sri Lanka

5.1 Evolution of Network and Spectrum Management Infrastructure

The limited radio spectrum imposes severe restrictions on the capacity and pricing of future wireless communication systems, necessitating new technologies and synchronized global standards [1–7]. The evolution from 2G's frequency division multiple access (FDMA) to 3G's wideband code division multiple access (WCDMA) and 4G's dynamic spectrum sharing (DSS) highlights its growing importance [8–10]. The exploration of software-defined networks (SDNs) for spectrum management further underscores this trend [11]. Future 6G spectrum management requires innovative solutions like machine learning and dynamic spectrum sharing, integrating various frequency bands, and harmonizing spectrum distribution through international cooperation and regulatory frameworks [12].

5.1.1 Wired and Wireless Communication Systems

Wired communication systems, ranging from traditional systems like plain old telephone systems (POTS) to contemporary systems like asymmetric digital subscriber lines (ADSL) and very-high-speed DSL (VDSL), utilize diverse spectrum management architecture and approaches [13]. The POTS, an early wired communication system, did not employ any spectrum management methods due to its narrow band nature [14]. The regulation of these channels was often overseen by recognized regulatory agencies. Frequency division multiplexing (FDM) is a key spectrum management technique utilized for ADSL and VDSL [15]. In this context, the spectrum is partitioned into distinct frequency bands that are allocated

Intelligent Spectrum Management: Towards 6G, First Edition.
Edited by Sridhar Iyer, Anshuman Kalla, Onel Alcaraz López, and Chamitha De Alwis.

to users in a dynamic manner. The ADSL and VDSL systems utilize a distinct strategy for spectrum management known as user profiling and service selection. This strategy guarantees that users are categorized and assigned priority based on their profile. This guarantees optimal utilization of the spectrum.

Bluetooth and Wi-Fi technologies are widely used in modern applications for wireless communication systems. Bluetooth employs frequency hopping spread spectrum (FHSS) technology, which divides the available spectrum into narrow-band channels. This enables various devices to switch to multiple available channels during transmission based on their usage and demand. Adaptive frequency hopping (AFH) is a sophisticated version of FHSS in which the hopping patterns are selected according to the level of activity on each channel. Wi-Fi mostly utilizes orthogonal frequency division multiplexing (OFDM) and adheres to regulations set by authorized entities to effectively control the spectrum [16, 17]. The OFDM partitions the available spectrum into distinct subcarriers, enabling simultaneous transmission of data across these subcarriers.

5.1.2 Requirement of Advanced Spectrum Management Infrastructure

After having explored the various forms of legacy spectrum management infrastructure, it is crucial to comprehend the increasing demand for adaptable and dependable spectrum management infrastructure. The need is a significant driving force that necessitates effective spectrum management. The proliferation of numerous devices resulting from the swift implementation of Internet of Things (IoT) and the rapid expansion of short-range wireless communication systems has caused a rising need for spectrum allocation [18]. Therefore, the presence of a well-functioning spectrum management infrastructure is essential. The rising demand has been further fueled by advancements in electronic technology. As the design and fabrication of integrated circuits have improved, communication systems have been able to accommodate more power and memory. This has resulted in enhanced performance and increased complexity. Ultimately, the existing allocation rules and practices have resulted in higher demand for spectrum, as the current processes have restricted access to it. However, the primary cause for these processes can be attributed to historical factors and the lack of synchronized policies implemented by many nations and institutions.

So far we have discussed the evolution of the legacy wired and wireless network and spectrum management infrastructure. Next, we will focus our attention toward the modern solutions for spectrum management, which involves the network virtualization and SDN technologies.

5.2 Network Virtualization—Toward Software-Defined Networks

The emergence of 5G networks necessitates innovative and adaptable design paradigms in telecommunications to support various services like broadband, IoT, and analytical machine-type connections. Network virtualization, pivotal in this transition, enables software-based emulation of hardware components for enhanced adaptability, resource utilization, and managing the complexity of 5G networks. Softwarization, which is a fundamental aspect of virtualization, assists in the accurate control of radio resources, coordinating the distribution of spectrum, and guaranteeing the efficient utilization of spectrum in radio access networks (RANs) [18–22]. SDNs complement this by separating control and data planes, allowing centralized management and flexible configuration. The integration of network function virtualization (NFV) with SDN further virtualizes network functions, enhancing efficiency and adaptability in 5G environments [19–21]. This chapter sets out to explore the foundational concepts of network virtualization and SDN, examining their historical development, core components, challenges, and future prospects within 5G networks.

5.2.1 Foundations of Network Virtualization

Network virtualization enables the simultaneous existence of several virtual networks on a shared physical infrastructure. In a network virtualization environment (NVE), a virtual network consists of virtual nodes and virtual links [23]. In essence, a virtual network is a portion of the resources within the underlying physical network. The goal of network virtualization is to separate functionalities in the networking environment by dividing the roles of traditional Internet service providers (ISPs) into infrastructure providers (InPs) and service providers (SPs). The InPs are responsible for overseeing the physical infrastructure, whereas SPs establish virtual networks by consolidating resources from numerous InPs, hence providing comprehensive network services from start to finish [24].

The concept of numerous concurrent networks is not new and has been extensively examined in networking literature in many contexts. The studies have focused on notions such as virtual local area networks (VLANs), virtual private networks (VPNs), and overlay networks [24]. Each of these notions has contributed, in various ways, to the advancement and comprehension of network virtualization. The motivation behind the push for network virtualization is to combat the stagnation of the Internet and encourage creativity and progress in the field.

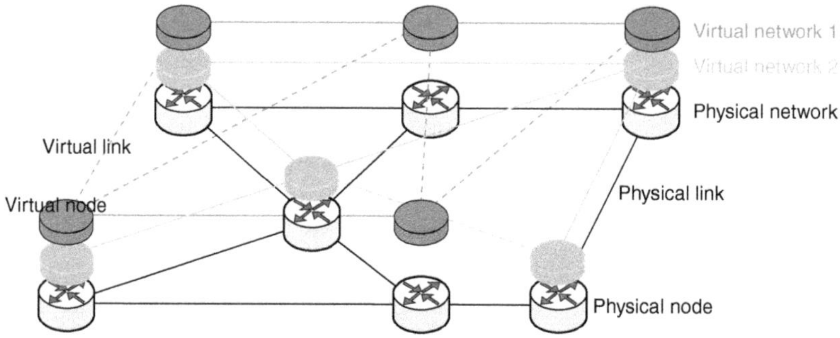

Figure 5.1 Basic components of network virtualization. Source: [24]/Springer Nature.

Expanding upon the fundamental comprehension of network virtualization, it is crucial to further explore the complex architecture that supports it. The majority of studies, as depicted in Figure 5.1, clarify the importance of a traditional virtual network topology that functions on top of a shared infrastructure. This architecture is based on two main components: the virtual node and the virtual link [25]. In a highly adaptable network virtualization setting, each fundamental node is constructed to contain many virtual nodes and support multiple virtual links simultaneously [27].

The architecture (Figure 5.1), in general, suggests a one-to-one relationship between the virtual nodes of a network and its corresponding physical entities. In contrast to this, link virtualization needs to realize a mapping of virtual links between a physical link and must guarantee link bandwidth for each virtual link. Thus, due to this difference between physical and virtual network topology, a virtual link may extend beyond two physical nodes [24]. A sophisticated viewpoint indicates that these virtual nodes frequently appear as virtual routers, with data plane virtualization developing as their essential component [28]. Initial investigations into node virtualization primarily focused on the complexities of isolating, modifying, and dividing fundamental hardware resources. This entailed complex manipulation of elements such as the CPU, memory, and input/output (IO) bandwidth, supported by developments in paradigms like para-virtualization (as demonstrated by platforms like XEN [29]) and operating system virtualization (such as OpenVZ [27]).

Virtual links play a crucial role in the design by connecting various virtual nodes to replicate the structure of the virtual network. Mapping virtual links to physical ones with precision is crucial in network virtualization, while also assuring the preservation of each virtual link's bandwidth [24]. An interesting finding is the potential difference in the structures of physical and virtual networks, resulting in virtual connections that involve more than two physical nodes, known as multihop virtual linkages [24]. Due to the inherent intricacy of these arrangements, every physical node in a virtual network must assume one of two roles: either as a link end node, which terminates the virtual link, or as a link intermediate node, which acts as a conduit. The implementation options for link virtualization have been a subject of disagreement due to various concerns such as standardization challenges, security considerations, resource allocation, and performance concerns. Nevertheless, there is a general agreement regarding the possibilities of technologies such as time-division multiplexing (TDM) and wavelength division multiplexing (WDM) for the purpose of isolating and dividing bandwidth. Modern networking technologies, such as multiprotocol label switching (MPLS) and asynchronous transfer mode (ATM), offer a strong basis for creating virtual links.

The Internet has experienced substantial changes since its creation. The landscape is undergoing significant transformation as new technology and innovative methods of leveraging network services are introduced. Nevertheless, the existing architecture of the Internet is not devoid of its constraints [29]. As emphasized in [24], the Internet's structure has been burdened with several difficulties, such as problems related to its ability to handle large amounts of data, worries regarding its security, and the requirement for effective data routing. These issues have compelled researchers and industry experts to reassess the fundamental concepts of the Internet.

Network virtualization has notable advantages, particularly in terms of its flexibility. By enabling the coexistence of several diverse network topologies, it creates an environment that promotes diversity and offers improved management. In addition, the isolation provided across coexisting virtual networks enhances fault tolerance, security, and privacy [24].

The development and understanding of network virtualization have been impacted by various preexisting notions. The mentioned networks encompass programmable networks, VLANs, VPNs, and overlay networks [18]. Each of these factors has influenced the development of network virtualization, adding to its specific features and capabilities.

Although network virtualization presents various benefits, it also presents significant difficulties. The implementation and maintenance of this system are characterized by complexity, potential performance overheads, and concerns regarding compatibility and interoperability. These challenges must be resolved. Nevertheless, as research and development in this field continue, it is expected that answers to these difficulties will arise, thus strengthening the significance of network virtualization in the future of networking [24].

5.2.2 Role of SDN in Network Virtualization

The SDN originated from earlier initiatives such as Epsilon's switch management protocol in 1998 and the Ethane project [25]. It is crucial to differentiate between

SDN and NFV. The SDN primarily involves separating control and data planes, whereas NFV involves virtualizing physical networking equipment. Both SDN and NFV are significant advancements in network innovation. The strength of SDN resides in its software-centric methodology, providing swift implementation and upgrades on contemporary multicore computers. The flexibility demonstrated in applications such as Google's VAHDAT project enables the most efficient use of network resources and the prioritizing of essential data flow. In addition to this, SDN's past and current progress, such as OpenFlow and NOX, highlight the importance of open standards, interoperability, and innovation, establishing SDN as a fundamental aspect of modern networking.

Conventional networking models are filled with intricate complications. Traditional devices, which have several layers ranging from management to data plane, frequently experience limitations in terms of visibility and control. This complexity is worsened when human interventions, which are usually responsive rather than proactive, become the main method of network changes, as seen in Figure 5.2. Introducing SDN is an innovative method designed to address these difficulties. The controller is a crucial component of the SDN architecture, serving as an intermediate between human instructions and machine operations. As depicted in Figure 5.2, the controller has the capability to swiftly adjust in real-time, actively rerouting flows and updating the flow tables of devices. The SDN controller is centered around the crucial North and South bound interfaces. The Northbound interface ascends, establishing a connection

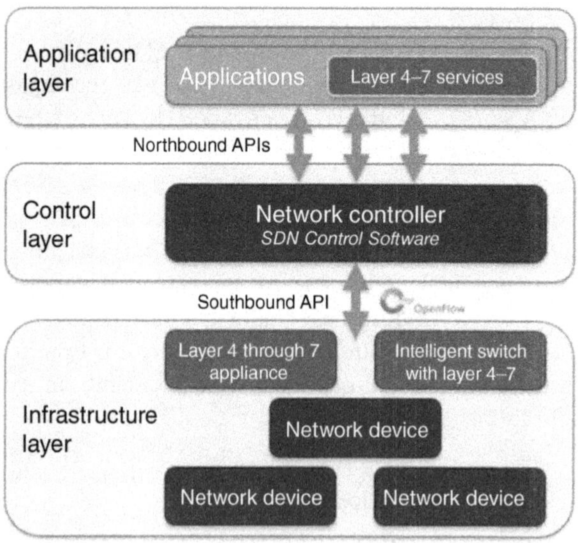

Figure 5.2 SDN architecture. Source: Adapted from [26].

with the management plane, while the Southbound interface descends, facilitating communication with the data plane. The East/West interface enables smooth communication between logical controllers, improving network coherence. Curiously, whereas the interface that handles data flowing from the South has achieved standardization through protocols like OpenFlow, the interface that handles data flowing from the North is still lacking such uniformity. Nevertheless, there are ongoing endeavors by groups such as the "North-bound application programming interface (API) WG" to provide APIs that encompass the intricacies of the north-bound management plane, thereby separating it from the SDN controller platform [26].

The SDN fundamentally advocates for centralized control of the data plane. The SDN centralizes network intelligence in the controller or control plane, creating an environment conducive to developing advanced applications that enhance network performance. In brief, the SDN architecture can be divided into three main layers: the application, control, and data layers, each with a specific function in coordinating a more efficient and agile network environment. The current networking environment is characterized by the rapid integration of SDN and NFV. These technologies, although different, work together smoothly, bringing about a period of programmability, automation, and scalability. Network services currently benefit from unparalleled flexibility, creating opportunities for groundbreaking solutions that are beyond conventional limitations.

5.2.3 NFV: Complementing SDN in Network Virtualization

Network slicing is a significant innovation that has emerged from the combination of SDN and NFV. Network slicing enhances resource consumption by enabling the construction of many virtual networks on a single physical infrastructure. Service chaining enhances this integration by dynamically combining virtual network functions (VNFs) to provide complete services. Efficiently managing and distributing resources across many domains necessitates careful coordination of numerous services.

5.2.4 Challenges in SDN-Driven Network Virtualization

Nevertheless, similar to many emerging technologies, the convergence of SDN and NFV is not without its hurdles. There are significant worries over security, as well as a strong desire for interoperability. The ongoing effort to achieve standards is essential for establishing a unified networking ecosystem. Although performance measurements show potential, they require more refinement, and the primary focus on reliability remains of utmost importance [30].

5.2.5 Emerging Trends and Developments

Further exploration of SDN and network virtualization reveals a multitude of advancements. The study in [31] provides a comprehensive and broad examination of the field, discussing important areas of research. These involve precise measurements and designs of SDN network hypervisors, complex mathematical models specifically for hypervisor locations, and algorithms customized for virtual network provisioning. Tools like "perfbench" [32, 33], a remarkable benchmarking tool, and designs such as "HyperFlex" [34–36] demonstrate the significant progress achieved in the field. Optimization methods, such as those based on mixed integer linear programming (MILP), aim to reduce the delay in the control plane. On the other hand, systems like "NeuroViNE" and "o'zapft is" utilize machine learning (ML) techniques to improve the efficiency of algorithms. The incorporation of SDN and network virtualization signifies a significant advancement in the field of networking. As we progress on this journey, it is crucial to consistently improve, investigate, and create new ideas, guaranteeing that we tackle the difficulties and completely achieve the vast possibilities of this integration.

5.3 Network Slicing: A Pillar for Spectrum Management in Modern Networks

Network Slicing is a cutting-edge technique in the contemporary telecommunications field, that distributes the computation, storage, and communication resources of a network to different services, whereby they can offer guaranteed performance levels. It allows different services to coexist in the same network infrastructure, preparing the path for the future generation of wireless networks.

5.3.1 Definition and Overview of Network Slicing

Network slicing, a prominent feature in future 5G networks, involves creating distinct virtual networks, or "network slices," over a shared physical network infrastructure. These slices, logically separate and spanning various domains and technologies, are essential for efficient resource allocation. Addressing challenges in contemporary networks, network slicing offers several benefits such as:

1. **Service Isolation:** It enables multiple services to coexist within a shared network without affecting each other's performance, thus ensuring service segregation.
2. **Performance Guarantees:** By efficiently distributing resources, network slicing ensures that each service meets its specific performance metrics.

3. **Heterogeneous Service Support:** To cater to the growing demand for diverse services, network slicing allocates RAN resources to support a variety of service requirements.
4. **Efficient Resource Utilization:** The adoption of nonorthogonal network slicing has led to substantial improvements in network performance, enhancing the overall resource utilization.

Recent research [38–45] has focused on the development, challenges, and solutions of network slicing across different sectors like radio, edge, transport, and core networks. The integration of multiaccess edge computing with network slicing marks a significant evolution, offering customized and high-speed services. This has led to a consensus on adopting a cloud-native 5G microservices architecture, particularly beneficial for network slicing facilitated by multiaccess edge computing.

5.3.2 Network Slicing: Components and Types

Network slicing is created to accommodate a wide range of different services within one network, it efficiently coordinates several functions while guaranteeing the best use of resources. However, what is the fundamental basis for this extraordinary ability? In what follows, we explore the complex elements and diverse categories of network slicing.

The network slicing design is a remarkable feat of engineering, integrating numerous layers to create a unified structure (see Figure 5.3). The service instance layer is situated at the highest point. It transcends being merely a layer and instead represents a domain where services attain their actual meaning. This layer guarantees the recognition, comprehension, and prioritization of any service, whether it is a high-definition video stream or a crucial piece of data from a remote sensor. The network slice instance layer connects the service requirements with the available resources. Consider it as the conductor of a symphony, orchestrating the different instruments (in this scenario, resources) to generate a harmonious output customized for each service. The resource layer serves as the fundamental basis for this architecture. The raw potential is found in the form of computation, storage, and communication resources. The bedrock layer guarantees the availability, accessibility, and optimal utilization of resources [42].

The appeal of network slicing lies in its precise level of detail, which is accomplished through its diverse components [41–43].

- **Network Slice Instance (NSI):** Envisage an end-to-end logical network, where resources, functions, and connectivity coalesce. This is an NSI which is a holistic network entity, ready to serve.

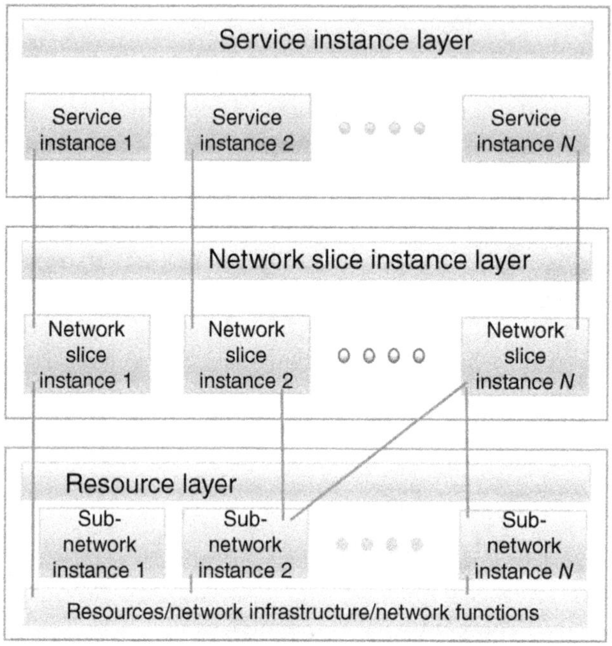

Figure 5.3 Network slicing layers. Source: [42]/CC BY 4.0/IEEE.

- **Network Slice Type:** In addition to resources and functions, there is a recognition of purpose, which is represented by the network slice type. This type classifies NSIs according to broad service categories, thereby encapsulating their intended use.
- **Network Slice Template (NST):** Standardization is key in a vast network, and NSTs ensure this standardization. They are the blueprints, the molds that give shape to NSIs.
- **Network Slice Subnet Instances (NSSIs):** Upon further investigation, we find NSSIs, which are groups of virtual functions that combine to create components of an NSI. They provide an additional level of detail, guaranteeing that each component of the network is exactly customized.
- **RAN, Edge, Transport, and Core Slices:** These are the fundamental slices, with each one responsible for a distinct section of the network. They oversee the management of radio access in RAN slices and ensure the proper functioning of essential network services in essential slices, ensuring comprehensive coverage of the whole network.

Diversity is essential for the flourishing of the digital world. The demands placed on modern networks are extensive and diverse, ranging from basic text messaging to real-time autonomous vehicle management. Network slicing acknowledges the variety of needs and preferences by implementing customized segments, each optimized for distinct service demands:

- **Ultrareliable Low Latency Communications (URLLC) Slices [39, 40]:** The domain of URLLC slices is characterized by its criticality and immediacy. Envision a where a surgeon does a surgical procedure from a considerable distance of thousands of miles. Even the slightest delay or temporary interruption in data transfer can result in severe effects. The URLLC slices are specifically intended for applications that are crucial for the success of a mission. These entities prioritize not only velocity, but also steadfastness and dependability. In addition to medical applications, they are utilized in areas such as autonomous transportation, where a minimal delay in data transmission might determine the outcome between safe navigation and an accident. These slices represent the highest level of accuracy, guaranteeing that data is transmitted with constant uniformity.
- **Enhanced Mobile Broadband (eMBB) Slices [39, 44]:** The eMBB slices are relevant in a time where entertainment and immersive experiences are highly emphasized. Imagine a virtual reality concert where numerous users concurrently partake in a live performance with exceptionally high resolution. These experiences require the transfer of large volumes of data at extremely fast speeds. The eMBB slices address the need for high data transfer rates, guaranteeing a smooth and uninterrupted experience when engaging in activities such as streaming a 4K movie or playing an augmented reality game.
- **Massive Machine Type Communications (mMTC) Slices [44]:** As we approach a future where all things are interconnected, the mMTC slices play a crucial role. Imagine a futuristic city where various elements such as traffic lights, utility meters, street cameras, and even trash bins are networked and able to communicate with each other and centralized systems. The data requirements in this context are distinct – rather than prioritizing fast speeds, the emphasis is on tolerating a large number of devices, each transmitting little packets of data. The mMTC slices are specifically designed to accommodate the extensive connectivity of large-scale devices, ensuring seamless operation of a vast network of IoT devices, hence enhancing urban living and streamlining industrial processes.

The fundamental appeal of network slicing resides in its inherent flexibility. By providing tailored solutions for diverse needs, it guarantees that all digital requirements, whether they include crucial real-time operations or immersive entertainment, are met with exceptional efficiency.

5.3.3 Benefits, Challenges, Threats, and Use Cases of Network Slicing

Telecommunication has experienced numerous advances, but few have been as revolutionary as network slicing. This concept, which is crucial for the future of connectivity, has the potential to fundamentally transform the way networks operate and provide services. In order to completely understand its significance, one must thoroughly explore the advantages it offers, the difficulties it presents, the potential patterns it indicates, and the multitude of practical applications it serves.

Benefits of Network Slicing

- **Resource Optimization [41, 49, 50]:** Network slicing's dynamic nature enables unmatched optimization of resources. It guarantees that spectrum resources, which were previously fixed and frequently not fully utilized, are now allocated flexibly according to real-time demand. This approach not only preserves important spectrum resources but also minimizes waste and inefficiencies that are inherent in fixed allocations.
- **Service Differentiation [38, 39, 41, 48]:** In a conventional network environment, various services frequently vie for the same resources, resulting in congestion and diminished service quality. Network slicing, by creating customized slices for individual services, guarantees that each application receives its optimum environment. This guarantees that a latency-sensitive application, such as real-time gaming, does not have any negative effects due to a bandwidth-intensive operation, such as video streaming, occurring simultaneously.
- **Enhanced Flexibility and Scalability [35, 43–45]:** Digital needs are dynamic, subject to change depending on factors such as time, events, or user behavior. The intrinsic versatility of network slicing allows for rapid scaling up of the network during periods of high demand and scaling down during periods of low demand, ensuring optimal resource use at all times.
- **Future-Ready Infrastructure [39, 41, 43]:** As the digital environment progresses, the services and applications it facilitates also advance. Network slicing guarantees the perpetual readiness of the network, whether it is to accommodate emerging applications or to adjust to novel data transmission methods.

Challenges Looming on the Horizon

- **Spectrum Fragmentation [38, 42, 47]:** Network slicing, while its agility, must function within the limitations of the spectrum resources that are now accessible. Allocating fragmented spectrum resources can be challenging, since it often necessitates the use of advanced algorithms to assure contiguous and optimum allocations.
- **Complexity in Implementation [38, 42, 46]:** The potential of network slicing is vast, but the complexity of its implementation is also significant. To achieve

smooth transitions, uninterrupted service throughout reconfigurations, and uphold the delicate equilibrium across various segments, a sophisticated network orchestration is necessary.

- **Security Concerns [38, 42, 44]:** As with any system that is subject to change, ensuring security becomes of utmost importance. Preserving the confidentiality of data within a slice and making sure that slices are impervious to breaches or external attacks introduces an additional level of intricacy to the network slicing paradigm.
- **Regulatory and Economic Impediments [38, 47, 49, 50]:** The process of allocating and deallocating resources in a dynamic manner presents distinct regulatory obstacles. Developing a framework that enables such adaptability, while guaranteeing equitable utilization and compliance with regulatory standards, poses a substantial difficulty. This is further compounded by the economic ramifications of such allocations, particularly in cases where spectrum is shared or traded.

Emerging Trends: The Future Beckons

- **Artificial Intelligence (AI) and Automation [44, 49, 50]:** Network slicing's complexity and dynamic nature make it an ideal contender for optimizations driven by AI. By employing predictive analytics, AI can proactively distribute resources, guaranteeing that the network remains consistently ahead of the demand.
- **Integration with Other Technologies [38, 39, 44]:** The integration of technologies such as IoT, blockchain, and edge computing with network slicing indicates a future in which networks are not just responsive but also decentralized and safe.
- **New Spectral Frontiers [39, 47]:** The relentless pursuit of increased bandwidth and enhanced communication is challenging the limitations of conventional frequency bands. By combining network slicing with the investigation of sub-6 GHz, millimeter-wave (mm-wave), and terahertz (THz) bands, we can expect to discover new possibilities in communication.

Diverse Use Cases: The Spectrum of Possibilities [38, 39, 42, 44]

- **Healthcare Revolution:** Network slicing has the potential to transform healthcare by enabling telemedicine consultations that connect urban and rural areas and facilitating real-time remote procedures. This technology can enhance accessibility and efficiency in the healthcare sector.
- **Autonomous Transportation:** Autonomy is the key to the future of transportation. Network slicing is crucial for autonomous vehicles, drones, and other forms of transportation that depend on continuous data transmission, as it offers dependable and instantaneous communication.
- **Smart Infrastructures:** Network slicing is a fundamental component of intelligent infrastructures, such as smart cities that adjust to the needs of their

residents or smart grids that optimize electricity distribution according to real-time demand.

- **Entertainment and Immersive Experiences:** Virtual reality (VR) and augmented reality (AR) are revolutionizing various sectors, including entertainment, education, and industries such as real estate. Network slicing proves to be a valuable asset for these immersive experiences, since they heavily rely on fast and reliable internet connections with large data capacity and minimal delay.
- **Industrial Automation and IoT:** The factories of the future are characterized by interconnectivity, where machines, products, and raw materials communicate with each other in real time. Network slicing guarantees the smooth operation of this complex network of communication, marking the beginning of the Industry 4.0 era.

Network slicing is a highly influential factor in the realm of telecommunication advancements. It envisions a future in which networks are not merely channels for data, but rather intelligent entities that can adjust, anticipate, and optimize. As we approach this shift, it is evident that network slicing, with its numerous advantages and extensive possibilities, will have a crucial impact on molding the future digital world.

5.3.4 Looking into Future of Network Slicing

Within the ever-evolving realm of telecommunications, the notion of network slicing emerges as a prominent concept, offering the potential for a future in which each unit of data and every service can efficiently navigate their most advantageous route. As we look toward the horizon, it becomes clear that the combination of network slicing with other technological breakthroughs will profoundly influence the structure of future communication networks.

Network slicing, however impressive, is closely intertwined with emerging technologies that enhance its importance. The emergence of network virtualization and SDN creates the conditions for the full realization of network slicing. Network virtualization enables the separation of resources from physical infrastructures, allowing for the creation of multiple distinct logical networks using a single physical backbone. However, SDN enhances programmability by separating the control and data aspects of the network. This, when paired with network slicing, allows for dynamic adaptation to the real-time requirements of various services.

As we move toward more sophisticated wireless systems, the importance of network slicing becomes increasingly crucial. The potential of 5G and future generations of wireless technology extends beyond simply faster speeds. It encompasses a wide range of applications, including the development of interconnected smart cities and the establishment of communication networks for autonomous vehicles. Network slicing serves as the foundation, providing each application with its own distinct portion of the spectrum that is customized to meet its specific requirements. However, this advancement also invites the investigation of unexplored ranges of electromagnetic radiation. Conventional frequency ranges, which are increasingly becoming fully utilized, are driving the advancement of technology into higher frequency bands such as mm-waves, THz frequencies, and even visible light communication. Network slicing will be essential in these unexplored areas, as it will guarantee the most effective and efficient use of resources.

The impact of network slicing extends beyond technology and has far-reaching consequences in the business domain. Providing tailored network solutions enables the development of inventive business models. Operators may now provide organizations with customized network environments, which guarantee both efficiency and cost savings, instead of offering a generic solution. For the end user, this means they will have an exceptional service experience, regardless of whether they are streaming multimedia, participating in real-time gaming, or accessing crucial real-time data.

Nevertheless, network slicing is not exempt from the difficulties that typically come with innovation. The prospect of dynamically optimizing slices for a range of services introduces intricacies. Overcoming these obstacles involves the design of these segments, guaranteeing the ability to adjust in real time, and managing limited spectrum resources. The future, however, holds great promise. The future network slicing solutions will utilize AI and ML to predictively optimize resources in advance. In addition, with the advent of 5G technology, cloud-native innovations are poised to revolutionize the concept of network slicing. Network slicing, which incorporates microservices, containers, and orchestration, is becoming highly efficient and naturally scalable, capable of meeting the dynamic demands of the future [46, 47].

Ultimately, as we find ourselves on the verge of a digital revolution, network slicing emerges as the fundamental building block. It offers a vision of a future in which networks are flexible, services are optimized, and the spectrum is fully utilized. Network slicing will serve as the connecting element that integrates many technologies into a cohesive, efficient, and forward-looking system as their boundaries become less distinct.

5.4 Network Virtualization and Network Slicing for Efficient Spectrum Management

The swift advancement of wireless communication networks, particularly with the introduction of 5G and succeeding technologies, has presented numerous difficulties and opportunities. The key issue in addressing these difficulties lies in the effective administration of spectrum resources, which are limited but essential for maintaining uninterrupted communication. Amidst these problems, researchers and industry experts have developed several new tactics and technologies, with network virtualization and slicing taking the lead. They enable telecom operators to meet the growing demands for high-speed, reliable connectivity in an increasingly connected world.

Network virtualization and network slicing are pivotal in managing spectrum in modern telecommunications. Network virtualization involves creating virtual versions of network resources, allowing multiple virtual networks to operate on the same physical infrastructure. This approach increases efficiency and flexibility in managing network resources, including the spectrum.

Network slicing takes this a step further by segmenting a single physical network into multiple, distinct virtual networks (slices). Each slice can be tailored to meet the specific needs of different applications or services. This customization is particularly beneficial for 5G networks, which must support a wide range of services with varying requirements, from high bandwidth applications to low-latency IoT devices.

By using network virtualization and slicing, operators can allocate spectrum resources more dynamically and efficiently. These technologies allow for the spectrum to be allocated on an as-needed basis, adapting to real-time demands of the network. This dynamic approach to spectrum management is crucial for maximizing the utility of limited spectrum resources and catering to the diverse needs of modern digital services. The integration of these technologies in spectrum management leads to improved network performance, enhanced user experience, and more effective utilization of the limited spectrum.

The authors of [45] present a new architecture that effectively combines network slicing with spectrum management. The objective of this framework is to enhance the efficiency of radio access networks by effectively managing spectrum resources while maintaining high service quality standards. Integrative techniques emphasize the need to align different network components in order to attain overall efficiency.

The work [46] explores the intricacies of network virtualization, with a particular focus on its use in coordinating network resources. The report does not offer specific techniques for spectrum management, but it emphasizes the crucial importance of virtualization in defining upcoming wireless networks.

In [47], the authors present a novel approach by introducing a distributed system for allocating resources. Decentralized solutions are crucial for effectively managing spectrum resources in dynamic network contexts, where adaptability and scalability are of utmost importance.

The significance of integrating spectrum management techniques with service demands is emphasized by both [48] and [49]. They support methods that prioritize both efficiency and adaptability to the changing spectrum conditions and diverse service requirements.

Fundamentally, the field of spectrum management is currently experiencing a significant change, with technologies like as network virtualization and slicing playing a crucial role. As we explore this picture further, it becomes clear that integrating these technologies with spectrum management measures is not only desirable but necessary.

5.4.1 Integration of Network Virtualization and Slicing with Spectrum Management

Network virtualization (NV) and network slicing are crucial technologies that have emerged alongside the advent of 5G and future generations of networks. Ensuring the smooth incorporation of these methods into spectrum management is crucial for realizing the complete capabilities of future networks.

The authors in [45] explore the incorporation of network slicing with spectrum management. They hypothesize that the seamless integration of these components is vital for optimal RAN functioning. The integration facilitates efficient usage of resources, leading to improved service delivery to end users.

A different study, [48], highlights the necessity of adopting a service-oriented approach to spectrum management. This method can improve the flexibility of networks in dynamic spectrum situations by utilizing the ideas of NV and slicing. Furthermore, although certain works, such as [50], do not specifically address the combination of NV and slicing, they emphasize the significance of effective resource allocation in systems with a high density of available frequencies. The implicit references highlight the underlying connection between NV, slicing, and spectrum management.

5.4.2 Spectrum Sharing, Allocation, and Dynamic Access: A Deeper Dive

Within the domain of wireless communication, the act of sharing and allocating spectrum is crucial in order to optimize the use of the limited frequency bands that are accessible. With the increasing prevalence of 5G and future systems, the focus on these tactics becomes even more prominent, particularly when contemplating the incorporation of network virtualization and slicing technologies.

For example, the authors in [45] provide insights on the significance of network virtualization in spectrum sharing. They contend that the combination of network slicing and virtualization can enhance efficient spectrum sharing, hence optimizing network performance. This implies that as networks grow more virtualized, their capacity to share spectrum resources in a flexible and efficient manner will improve, ultimately benefiting end users. Nevertheless, numerous articles, including [46], lack comprehensive and specific proposals for spectrum sharing and allocation. The absence of this information may suggest that although the significance of spectrum sharing and allocation is well recognized, the methods to implement them, particularly in the context of network virtualization and slicing, are still being actively studied.

Notably, [47] presents a distributed network slicing approach to allocate spectrum efficiently. Decentralized solutions have the potential to provide improved scalability and flexibility, particularly in network settings that are dynamic and varied.

The point raised by [48] is relevant: whereas spectrum has typically been perceived as a fixed resource, the actual situation is somewhat different. The ever-changing characteristics of the spectrum require techniques that are not only efficient but also flexible. The study proposes that adopting a service-oriented strategy can effectively address this issue by allocating spectrum according to the real-time demands of the network. Dynamic spectrum allocation (DSA) is crucial for contemporary wireless networks, particularly due to the growing need for bandwidth and the scarcity of spectrum resources. By integrating network virtualization with slicing, DSA has the potential to introduce a new age of highly efficient and responsive networks. Within this particular framework, [45] discusses the significance of virtualization in the realm of dynamic spectrum management. The article argues that virtualized networks possess inherent advantages in managing the requirements of dynamic spectrum settings due to their modular and adaptable nature.

The author of [48] strongly promotes the use of a service-oriented approach to spectrum management. The research highlights the significance of synchronizing spectrum access mechanisms with immediate service requirements, guaranteeing the optimal utilization of the spectrum and preserving user experiences without any compromise.

5.4.3 Use Cases: Virtualization and Slicing in Spectrum Management

The substantial study and discussion have focused on the revolutionary possibilities of network virtualization and slicing, particularly in the context of spectrum management. Through the analysis of findings from several studies, we may discern and comprehend the practical applications and consequences of these technologies in the actual world.

The usefulness of network slicing in managing the RAN for 5G is emphasized in [45]. An important consequence of this is the capacity to provide a wide range of services without any interruption or difficulty. RAN slicing is used in urban areas to cater to a wide range of user demands, such as high-definition video streaming and IoT device connections. It ensures that each service type is allocated a specialized frequency slice that is tuned to meet its individual requirements. The level of detail in the spectrum allocation not only assures optimal use but also guarantees users a uniform and superior user experience. Furthermore, in situations where there is a significant demand for network utilization or a large concentration of people, the capability of RAN slicing to allocate spectrum resources in a flexible and dynamic manner becomes extremely valuable. It functions as a protective barrier against network congestion, guaranteeing consistent service quality even under difficult circumstances. Exploring the core concept of platform virtualization, as proposed by the study [46], we discover numerous possible uses. An exemplary application is the administration of network heterogeneity. Contemporary urban areas frequently face the challenge of managing a combination of antiquated and contemporary network systems. Virtualization platforms can serve as a bridge, unifying these different networks and guaranteeing smooth user experiences. Moreover, in situations requiring swift enlargement of the network, particularly in developing markets or remote rural regions that are difficult to access, virtualization becomes the guiding light. Virtualization enables the rapid deployment of networks by utilizing existing infrastructures, thereby extending access to far areas without the need for substantial expenditures in physical equipment.

The article [48] advocates for a service-oriented strategy, which has distinct and exclusive applications. For crucial services like emergency services or telemedicine, where reliability and low delay are extremely important, adopting a service-oriented spectrum approach guarantees that these services receive first priority. They are allocated the majority portion of the spectrum, guaranteeing that during emergencies, communication stays unimpeded. In the case of platforms that rely on delivering material in real time, such as streaming services or online gaming, a service-oriented spectrum allocation may adjust and distribute resources according to immediate needs. This guarantees that customers encounter minimal delays and obtain the best possible performance. Finally, the concept of [49] suggests the importance of applications that are crucial for contemporary smart cities. As these cities become more dependent on a wide range of devices, such as traffic sensors and smart grids, efficient resource management becomes crucial in guaranteeing seamless communication between each device. In industrial settings, particularly with the increasing use of Industrial IoT, it is vital to provide effective and conflict-free allocation of spectrum in order to maintain smooth and uninterrupted operations.

Ultimately, the experiments presented here offer a captivating preview of a future in which spectrum management, enhanced by virtualization and slicing, can effectively tackle a wide range of practical obstacles, therefore enhancing the efficiency and responsiveness of wireless communications.

The integration of network virtualization and slicing, along with effective spectrum management, has the potential to significantly transform multiple industries, including those that are now gaining attention.

1. **Autonomous Vehicles:** With the increasing popularity of self-driving cars, their functionality will greatly depend on the transmission of real-time data. Network slicing enables the allocation of specific spectrum slices for autonomous vehicles, guaranteeing the prompt delivery of data regarding traffic conditions, pedestrian movements, and other vehicles. This information is vital for assuring safe navigation.
2. **AR and VR:** Both AR and VR require a significant amount of data transfer capacity and minimal delay. Network virtualization enables service providers to prioritize spectrum allocation for AR and VR devices, guaranteeing seamless and immersive experiences for customers.
3. **Smart Agriculture:** Within the era of precision agriculture, producers employ sensors to oversee soil moisture levels, meteorological conditions, and crop health. By employing network slicing, efficient spectrum management may be achieved, enabling real-time communication among various sensors. This, in turn, assists farmers in making prompt decisions and enhancing agricultural yields.
4. **Remote Surgery:** The implementation of telemedicine is now revolutionizing the healthcare industry. However, with the emergence of 5G and future advancements, it may become possible to conduct intricate medical procedures, such as surgery, from a remote location. For such important applications, it is necessary to reserve certain portions of the radio spectrum allocated exclusively for their use. This allocation is ensured through network slicing, which guarantees that there will be no delay and the highest level of dependability.
5. **Smart Grids:** Smart grids will be crucial in optimizing energy use as we transition to more sustainable energy options. Network virtualization will greatly help these grids, which are equipped with sensors and IoT devices, by ensuring they have access to the required spectrum resources for optimal operation.
6. **Large-scale IoT Deployments:** As cities and industry increasingly incorporate IoT devices into their infrastructures, the available frequency range will grow crowded. Network slicing guarantees that these devices, regardless of whether they are located in a factory or are part of a city's infrastructure, are allocated specific portions of the wireless spectrum, which ensures seamless operations.

7. **Disaster Response:** Following natural disasters, effective communication becomes of utmost importance. Network slicing enables the allocation of exclusive spectrum resources for rescue operations, guaranteeing the coordination and effectiveness of relief efforts.

Essentially, as technology advances and becomes more deeply integrated into our lives, the requirements for the spectrum will increase dramatically. Network virtualization and slicing will be essential in assuring efficient utilization of this spectrum, accommodating both existing and emerging use cases.

5.4.4 Quality of Service and Quality of Experience Improvements Through Virtualization and Slicing

In the dynamic field of wireless communications, the pursuit of improving Quality of Service (QoS) and Quality of Experience (QoE) is of utmost importance. The use of technologies such as network virtualization and slicing provides innovative opportunities to accomplish these goals, and numerous research papers [45–48, 51–54] offer fascinating perspectives on this process.

The study [45] emphasizes the significant capacity of network slicing in attaining maximum efficiency in resource allocation. By customizing certain segments of the network to accommodate different service requirements, the occurrence of congestion is minimized, the delay is decreased, and the consistency of bandwidth is preserved. The network's dynamic flexibility allows it to respond to changes in user demands and competing services by modifying in real time to maintain a high QoS [45, 52]. Nevertheless, although the prospect of reliable performance is enticing, the challenges associated with overseeing many slices must not be disregarded. The inclusion of slice management may present difficulties, potentially impacting the overall effectiveness of the network.

Upon further examination, the core concept of [48] advocates for a distinct paradigm: an approach to spectrum management that prioritizes service-centricity. This methodology promotes optimal utilization of spectrum by aligning spectrum allocation with the specific demands of different services, hence boosting user experiences. For example, essential services that require fast response times might be given higher priority, guaranteeing that even during periods of heavy network traffic, their performance will not be impaired. However, adopting a service-oriented strategy presents its own unique difficulties. Classifying and ranking services, particularly in a constantly evolving environment where new applications and services regularly arise, can be overwhelming. Additionally, there is a potential danger of misallocating resources if the demands of the service are not appropriately assessed.

Another fascinating viewpoint arises from the issue discussed in [47]. Decentralization fundamentally guarantees scalability. By decentralizing resource allocation duties across the network, the occurrence of bottlenecks is avoided, hence guaranteeing consistent performance as the network expands. Moreover, implementing a distributed strategy inherently minimizes the risk of single points of failure, hence enhancing the resilience of the network. Nevertheless, the difficulties in coordinating distant nodes and the possible discrepancies in resource allocation regulations may impede the complete realization of decentralization's capabilities.

Upon analyzing these observations, it becomes clear that the journey toward improving QoS and QoE in the era of 5G and beyond is filled with both favorable circumstances and difficulties. Network virtualization and slicing offer a promising basis due to their inherent capabilities. However, the effectiveness of these technologies in actually changing wireless communications will depend on the approaches and strategies employed.

5.4.5 Security in Virtualized and Sliced Networks: Spectrum Management's New Frontier

The combination of network virtualization, slicing, and spectrum management has created a new and exceptional era in wireless communications. Although this convergence offers exceptional efficiency and agility, it also brings up a new set of security issues that are closely linked to the intricacies of spectrum management.

Spectrum, which is essential for wireless communication, has long been considered as an unchanging asset. Nevertheless, due to the emergence of dynamic spectrum management systems, it is now being regarded as a flexible entity that is assigned and reassigned according to real-time requirements. The dynamic nature of this system, although it ensures the most efficient use of the available spectrum, also brings forth possible security weaknesses. For example, a malevolent actor could try to manipulate the spectrum allocation procedures, resulting in interference or even the denial of services to authorized users.

The article [47] highlights the significant advantages of decentralization in the administration of spectrum. Nevertheless, the implementation of a decentralized spectrum allocation strategy, although it offers the advantage of flexibility, also brings about possible security vulnerabilities. Without a centralized control, maintaining consistent security measures over the entire range becomes problematic. Moreover, enemies can utilize the flexible redistribution of frequency bands in these systems to carry out advanced jamming or interference assaults, thereby disrupting lawful communications.

However, [48] introduces the idea of service-centric spectrum allocation. While this guarantees that essential services are given priority in the allocation of spectrum, it also poses a security quandary. Malicious actors may attempt to compromise the most vital communications by targeting the spectrum slices that are allotted to high-priority services.

In addition to the insights gained from the provided study, the incorporation of network slicing with spectrum management introduces an additional level of intricacy to the security environment. Every network slice, which is customized for particular services or user groups and assigned unique spectrum resources, becomes a potential objective. A malicious actor, through compromising the spectrum allocation algorithms of a certain portion, could cause disruption to its services, thereby impacting both the QoS and QoE for users inside that portion. Nevertheless, the challenges also offer prospects. The level of detail provided by network slicing, when combined with the ability to regulate spectrum dynamically, can be utilized to improve security. Through the process of isolating and implementing specific security protocols for each slice, flaws in the spectrum management methods of one slice may not necessarily pose a threat to others. Furthermore, the inherent real-time monitoring capabilities of dynamic spectrum management can be expanded to promptly identify anomalies or potential security breaches, enabling prompt remedial actions.

5.4.6 Regulatory and Policy Implications in Spectrum Management for Virtualized and Sliced Networks

The combination of network virtualization, slicing, and spectrum management is driving innovation in wireless communications. The technical gains are evident, but they also unavoidably present several regulatory and policy issues. It is essential to address these issues in order to ensure the strength and effectiveness of these advances, while also finding a harmonious equilibrium between technological ambitions and society requirements.

The precious and invaluable resource of spectrum has consistently been the focal point of regulatory disputes. In the current era, when spectrum is controlled, allotted, and reallocated depending on real-time demands, there is an increasing need for comprehensive and flexible rules. Conventional methods of assigning spectrum, which were fixed and predetermined, may no longer be adequate to handle the dynamic changes brought about by network slicing and virtualization.

Take into account, for example, the knowledge gained from the reference [48]. The paper advocates for a service-oriented approach to spectrum allocation, which aims to align spectrum resources with the specific requirements of various services. Although this strategy offers improved spectrum usage, it also presents regulatory obstacles. How can equity be guaranteed in the distribution of spectrum when certain services are given priority over others? What safeguards are implemented to deter monopolistic inclinations, wherein dominant services or providers may monopolize spectrum resources to the detriment of smaller competitors or emerging technologies?

In addition, the decentralized philosophy emphasized in [47] introduces its own set of policy considerations. The use of a decentralized spectrum management system, although it shows potential for improved scalability and agility, also gives rise to concerns of supervision and accountability. In the absence of a centralized authority governing spectrum allocations, how do regulators guarantee the prudent and ethical utilization of the spectrum? Who serves as the arbitrator and what criteria are used to resolve disputes or controversies regarding spectrum allocations?

In addition to the details of the given research, the incorporation of network slicing with dynamic spectrum management requires a reassessment of licensing models. Conventional licensing methods, typically characterized by long-term agreements and limited to specific locations, may require adaptation to more adaptable and perhaps instantaneous models to fit the dynamic nature brought about by network virtualization and slicing. Nevertheless, these obstacles also offer prospects. Regulators and policymakers have the opportunity to create policies that are not only adaptable to current technological improvements but also proactive in anticipating future innovations. By actively involving industrial stakeholders, researchers, and the public, a comprehensive framework may be created to protect the interests of all parties concerned. Implementing collaborative policymaking can guarantee that technology ambitions are guided by ethical, fair, and societal factors.

5.4.7 Future Trends: AI and ML in Spectrum Management for Virtualized and Sliced Networks

With rapid advancements in wireless communication technology, we catch exciting glimpses of what the future may include. The core of these futuristic concepts revolves around the use of AI and ML in spectrum management, particularly when examined in the context of network virtualization and slicing [49–54].

The AI and ML, due to their intrinsic ability to analyze large volumes of data and identify trends, have the potential to completely transform spectrum management. Within the framework of DSA, AI algorithms have the capability to forecast spectrum requirements by using historical data, user behavior, and external factors such as events or weather patterns. These predictive capabilities can secure the allocation of spectrum before any demand occurs, resulting in reduced latencies and improved user experiences [50, 52].

Although the potential of AI-driven spectrum management is vast, it is important to acknowledge that this field is still in its early phases of development. The complexity and dynamism of wireless networks, particularly when combined with network virtualization and slicing, pose a significant barrier in utilizing AI for spectrum management. Developing AI models capable of navigating this complex terrain and ensuring efficient spectrum allocation is a challenging endeavor. Moreover, as these AI models engage in real-time decision-making, the margin for mistake decreases. Incorrect spectrum allocations, even if temporary, have the potential to interrupt services, impacting both QoS and QoE.

The ML, which is a branch of AI, presents fascinating opportunities, particularly in the field of adaptive spectrum management. The ML algorithms have the ability to adjust and accommodate variations in network conditions by means of ongoing learning. Within the framework of network slicing, ML models have the capability to acquire knowledge about the spectrum requirements of various slices over a period of time. This enables them to enhance their allocation algorithms, hence achieving the most efficient utilization [52, 53]. The ability to adapt is particularly important in contemporary networks, which are defined by constantly changing user requirements and service environments.

However, the use of ML in spectrum management is not devoid of difficulties. A key issue is the level of openness in ML judgments. Understanding the reasoning behind the allocation decisions of ML models, particularly deep learning models, can be problematic because of their typically opaque nature, commonly referred to as "black boxes." The lack of transparency on this matter could present difficulties in terms of regulation and responsibility, particularly in situations where conflicts over spectrum allocation occur.

In addition to AI and ML, future developments in spectrum management suggest the incorporation of other developing technologies. Blockchain, with its decentralized and transparent nature, has the potential to be instrumental in guaranteeing equitable and tamper-resistant spectrum allocations. Edge computing can provide localized spectrum management, which involves making allocation choices closer to the users. This approach reduces latencies and improves responsiveness.

At this critical juncture of technological advancement, the outlook for spectrum management appears optimistic due to the integration of AI, ML, and other developing technologies [38, 39, 45, 54]. However, achieving this commitment necessitates a careful combination of creativity, forward-thinking regulation, and cooperative investigation. Through effectively addressing the difficulties and capitalizing on the advantages, we may form a forthcoming era in which the management of spectrum, supported by network virtualization and slicing, is both highly efficient and adaptable.

5.5 Spectrum Virtualization and Network Slicing Enabled Infrastructure for Spectrum Management

In the dynamic realm of telecommunications, the need for spectrum resources is currently at an unprecedented level. The emergence of 5G technology and the potential advancements it brings highlight the crucial importance of effectively utilizing the limited spectrum available. Spectrum virtualization is a significant innovation that is emerging in this context.

Spectrum virtualization is the process of converting real spectrum resources into virtual equivalents, which can be flexibly assigned to various services and operators. By adopting this approach, it not only tackles the issues related to limited spectrum availability but also optimizes its usage, guaranteeing that the spectrum resources at hand are utilized to their maximum capacity [46, 49, 50]. Spectrum virtualization is converting a fixed resource into a flexible one, capable of adjusting in real time to accommodate the changing requirements of networks and applications.

Various technologies have contributed to the implementation of spectrum virtualization. The SDN offers the required flexibility in routing and operations, whereas NFV enables the separation of network services from the hardware, resulting in more adaptability. Collectively, they establish the fundamental basis on which spectrum virtualization is constructed.

In addition, the utilization of modern methods like as OFDM and multiple-input multiple-output (MIMO) has significantly enhanced the potential of spectrum virtualization [54]. These techniques enable the generation of numerous logical channels from a single physical channel, hence increasing the overall capacity of a given spectrum.

Nevertheless, the process of spectrum virtualization is not without difficulties. As the spectrum is virtualized and divided among various entities, concerns regarding interference control, resource allocation, and QoS become prominent. Ensuring security in a virtualized system is a significant problem due to the possible vulnerabilities that may be exploited.

5.5.1 Infrastructure Requirements for Spectrum Virtualization

Spectrum virtualization is a cutting-edge technology in telecommunications that offers improved efficiency and flexibility. In order to fully achieve its potential, a strong and reliable infrastructure supported by various crucial elements is necessary.

The SDN/NFV framework plays a significant role in multiple studies on the subject [49, 50]. This architecture enables the dynamic setup and orchestration of network services and resources by separating the control plane from the data plane. The relevance of it is emphasized by the fact that it serves as the foundation for the implementation of spectrum virtualization. The framework enables the formation of coherent networks consisting of virtualized or physical network functions, interconnected smoothly. This architecture offers open programming interfaces, allowing external programs to access and engage with network slices.

The spectrum virtualization layer is an essential component of the infrastructure. The primary function of this layer is to establish many logical channels from a single physical channel. It utilizes modern technologies such as OFDM and MIMO to accomplish this conversion [50, 54]. By employing this approach, it maximizes the efficient capacity of a specific spectrum, guaranteeing the maximum exploitation of the resources at hand.

Network slicing is closely interconnected with the notion of spectrum virtualization. An essential component for the dynamic allocation of optimal resources is a dedicated network slicing layer. The primary function of this layer is to choose suitable segments for different services and users. The factors considered in making these judgments include a broad spectrum, including international mobile subscriber identity, traffic type, device mobility, and slice capacity. The complex process of decision-making highlights the significance of having a specialized layer that is only dedicated to network slicing.

The significance of a global controller in this design cannot be overstated. It performs a wide range of duties, including as authentication, traffic scheduling, and identification of possible threats like DDoS attacks. The global controller guarantees smooth operation and security of the virtualized spectrum by employing sophisticated techniques like PBKDF2, asymmetric queuing models, and Renyi entropy [50].

The integration of AI and ML into the infrastructure is unavoidable in the current day. The use of AI and ML approaches is becoming more common in order to facilitate autonomous and efficient decision-making for the operation and maintenance of network slices. AI and ML are transforming the management and operation of spectrum virtualization, encompassing tasks such as slice admission control, resource allocation, VNF deployment, and slice elasticity [49, 50].

The 5G core is becoming another fundamental component of the infrastructure. Functioning as the central hub, it actively distributes resource capacities throughout the network and adjusts the size and placement of workloads as needed. The dynamic nature of the 5G core guarantees the best utilization of spectrum resources at all times.

Moreover, channel state information offers important knowledge on the conditions of the wireless channel, enabling the strategic distribution of spectrum resources according to up-to-date information. This information becomes crucial, particularly in locations where wireless circumstances can change frequently.

Finally, the necessity for multidomain orchestration is apparent. To accommodate the expanding and diverse networks that encompass terrestrial, aerial, and satellite domains, it is necessary to have a concerted endeavor to establish network slices that cover the entire network and comply with service level agreements. This component has the capability to adjust network slices in response to changes in traffic and the development of network circumstances.

The SDN/NFV architecture is crucial to the discussion on infrastructure needs for spectrum virtualization. The control and data planes are separated by this fundamental element, enabling the flexible coordination of network operations and resources. In addition to this, there is the spectrum virtualization layer, which is specifically created to generate several logical channels from a single physical channel, utilizing advanced technologies like OFDM and MIMO. The network slicing layer plays a vital role in dynamically allocating resources and selecting slices based on various parameters [49, 50, 53, 54].

The global controller has a position of utmost significance, supervising the extensive range of the virtualized spectrum. It is responsible for tasks such as authentication, traffic scheduling, and watchful threat detection, making it the protector of the virtualized network. The AI and ML techniques, which are marvels of the modern day, play an active role in its progress rather than being passive observers [50, 53, 54]. They play an active role in advancing the system, facilitating autonomous and data-driven decision-making in the operation and administration of network slices. Enhancing the infrastructure are additional elements such as the 5G core, which efficiently assigns resources, and Channel State Information, providing immediate updates on wireless conditions. Finally, multidomain orchestration integrates the various terrestrial, aerial, and satellite networks into a unified system, guaranteeing smooth and uninterrupted network connections.

5.5.2 Virtualized Spectrum Management and Dynamic Spectrum Allocation in Modern Telecommunications

The management of spectrum resources is a crucial concern in the constantly changing field of telecommunications. As network needs increase, it becomes crucial to allocate and utilize these resources efficiently. Virtualized spectrum management (VSM) and DSA have emerged as promising solutions to address these difficulties. They offer flexibility, efficiency, and adaptability.

The VSM involves the process of converting physical spectrum resources into virtual ones, but there may be variations in its definition in different sources. This abstraction enables the dynamic allocation of these resources to different operators and services according to unique requirements. Although certain literature may not directly address VSM as an independent subject, it does allude

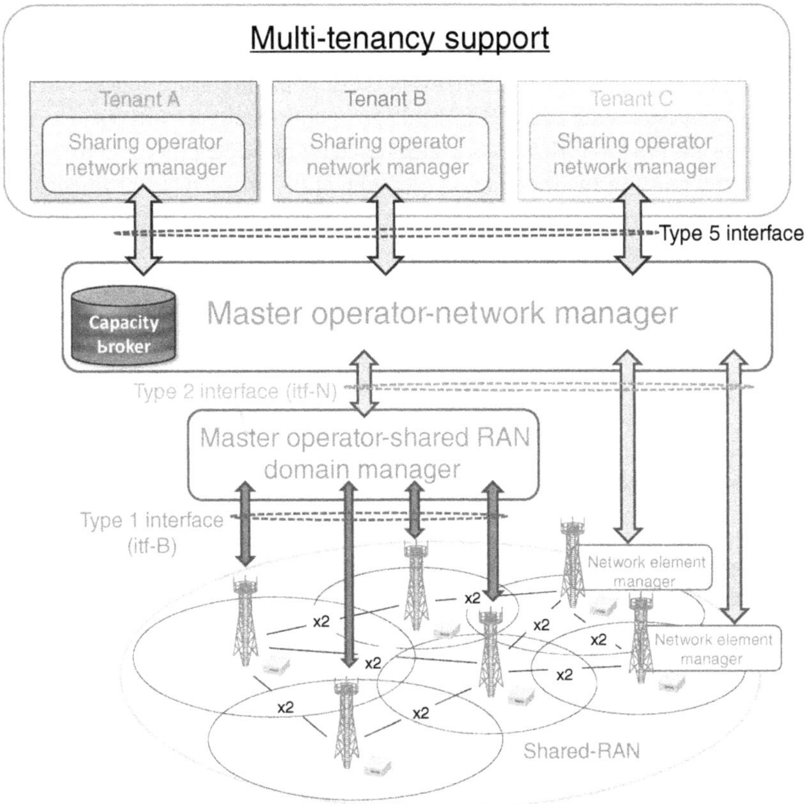

Figure 5.4 An architecture facilitating virtualized spectrum sharing. Source: [55]/IEEE.

to frameworks or approaches that are in line with its ideas. An example of an architecture suggested to facilitate virtualized spectrum sharing in multioperator 5G networks (see Figure 5.4) consists of three primary elements: a spectrum broker, a virtualization controller, and a resource manager [55]. The spectrum broker facilitates the coordination of spectrum sharing between many operators, ensuring that resources are assigned in accordance with specific requests. Simultaneously, the virtualization controller transforms physical resources into virtual ones, dividing the network to accommodate various services. The resource manager efficiently utilizes resources in each network slice, according to changing conditions and user needs.

The DSA enhances the VSM by guaranteeing that spectrum resources are not only virtualized but also allocated in an effective manner. The DSA, within the framework of spectrum virtualization, improves versatility by facilitating spectrum allocation across different entities, ranging from mobile network operators to service providers. The fundamental concept underlying DSA is to guarantee the allocation of spectrum resources in real time, taking into account the current network circumstances, requests, and availability. Some literature suggests utilizing neural networks, such as the HopFieldNet, to optimize spectrum resources by allocating resources for each network slice based on multiple criteria [56].

A recurring motif in the literature is the interaction between spectrum virtualization and other types of resource sharing, such as network virtualization and infrastructure virtualization. These notions jointly enhance the functionality of DSA by allowing the division of network resources into separate slices that are customized to support various services [48, 53].

Nevertheless, like every progressive technology, there are obstacles to overcome. A strong virtualization-based control and coordination structure is necessary to handle the intricate and varied spectrum-sharing scenarios in 5G networks. The pressing concern is to achieve scalability, dependability, and security in spectrum sharing and network virtualization.

5.5.3 Realizing the Future: Use Cases of Spectrum Virtualization and Network Slicing

The potential of spectrum virtualization and network slicing is immense, offering prospects to revolutionize the telecommunications industry in significant ways. However, the real evidence of their capability is demonstrated by their practical implementations—the specific scenarios where these concepts are put into action, transitioning them from theoretical ideas to actuality.

Smart cities offer highly appealing opportunities. The principles of spectrum virtualization and network slicing are well suited for these metropolitan landscapes, which are filled with interconnected systems and gadgets. Various services such as smart grids, transportation systems, healthcare, and security can be supported by distinct network slices, each designed with specific resources and security measures. The capacity to allocate spectrum resources in real time enables a smart grid to efficiently manage high energy demands during midday without disrupting a concurrent smart health monitoring system. The fine level of detail in allocating and managing resources guarantees that each service can function at its highest efficiency without any disruption.

Industry 4.0, the next phase in digitizing manufacturing, is a captivating field. The need for dependable communication in the intricate network of industrial processes is vital. Spectrum virtualization and network slicing address this by offering specialized slices, meeting demands like real-time data transmission for a robotic assembly line. Furthermore, these technologies are integral to Industry 5.0, in enhancing wireless communication. Spectrum virtualization optimizes usage by dynamically allocating frequencies, while network slicing creates isolated virtual networks tailored to diverse industrial needs. This synergy ensures efficient radio-frequency utilization, providing essential low-latency, high-reliability connectivity for real-time communication in smart factories. With agile spectrum management and the coexistence of varied services, spectrum virtualization and network slicing form a foundation for robust, adaptable, and scalable communication networks crucial for Industry 5.0's transformative advancements.

Vehicular networks offer a further intriguing application. Envision a future in which self-driving vehicles, unmanned aerial vehicles, and rapid transit systems seamlessly exchange information and collaborate with one another and the surrounding infrastructure. This situation necessitates distinct QoS prerequisites and patterns of mobility. A drone necessitating the delivery of a cargo may necessitate a slice with fast throughput and low latency for a brief period, but an autonomous car may require a more extended yet consistent connection. Virtualized spectrum infrastructures have the ability to adjust immediately, guaranteeing that each type of vehicle receives the necessary spectrum resources at the appropriate time.

Although the literature only presents a few selected use cases, the potential applications of spectrum virtualization and network slicing are extensive. It is possible to imagine future situations where remote surgeries become widespread, using specialized network slices to guarantee no delay. In a similar manner, extensive IoT networks that cover smart homes, wearables, and other devices can function in a coordinated way, ensuring that each type of device receives its fair share of the available frequency spectrum.

However, like any other promising technology, there are still existing problems. Although spectrum virtualization provides the advantage of flexible resource distribution, it requires strong management frameworks. Algorithms, perhaps utilizing sophisticated ML models, must be implemented to guarantee the most efficient allocation procedures. While these algorithms enhance the efficiency of spectrum utilization, they need to adjust to constantly changing user requirements and patterns of mobility. The ever-changing nature of these needs emphasizes the necessity for ongoing innovation in the industry.

5.5.4 Scalability and Flexibility in Spectrum Virtualization: A Deep Dive into Modern Telecommunication Needs

Spectrum virtualization, with its revolutionary capacity in telecommunications, encompasses more than just the abstraction of resources. The success of this depends on two crucial qualities: scalability and adaptability. With the increasing complexity of networks and the diversification of user needs, there is a critical need for scalable and flexible spectrum management solutions.

Scalability in spectrum virtualization refers to the ability to smoothly handle a growing number of users or jobs without sacrificing performance or efficiency. An ongoing topic in the research is the significance of dynamic and adaptive resource allocation in improving scalability. Systems can optimize performance during peak periods by customizing resource allocation based on service types and user needs, allowing them to respond to fluctuating network loads. The SDN and NFV technologies enhance scalability. They provide the adaptable and customizable administration of network resources, enabling systems to adapt to various applications and situations. The implementation of neural networks, such as the HopFieldNet, significantly improves resource allocation, strengthen network resilience, and address the issue of limited spectrum availability [56].

Flexibility emphasizes the ability of spectrum virtualization infrastructures to adapt. In the domain of 5G and beyond, where communication services exhibit varied and frequently demanding needs, adaptability is essential. Network slicing, as emphasized in several research, is a crucial element in this scenario [43, 45, 53, 54, 56]. Network slicing exemplifies the flexibility provided by spectrum virtualization through the implementation of dynamic resource allocation, optimal resource splitting, and the guarantee of service assurance and isolation. These capabilities are not merely theoretical concepts; they are currently being implemented in ongoing initiatives such as 5G-PPP, 5G-XHaul, and 5G-MoNArch.

Although scalability and flexibility are frequently praised as advantages of spectrum virtualization, they also present inherent difficulties. Creating frameworks that may effortlessly expand while maintaining optimal performance necessitates careful equilibrium. Likewise, guaranteeing adaptability while maintaining the security and a high standard of service requires strong management frameworks.

In anticipation of the future, with the increasing convergence of physical and digital realms, communications networks will face unparalleled pressures. Imagine a future when augmented reality applications operate in real time, billions of IoT devices exist, and autonomous vehicles communicate with both each other and the infrastructure. In a future like this, the ability to scale and adapt provided by spectrum virtualization would not just be wanted but necessary. Systems must efficiently distribute resources within microseconds, promptly respond to user demands, and maintain consistent performance under any workload.

5.5.5 Securing the Future: Challenges and Enablers in Spectrum Virtualization

Spectrum virtualization has evolved as a groundbreaking concept in the dynamic field of telecommunications, offering unparalleled efficiency and versatility. Nevertheless, along with this commitment, there are other obstacles to overcome, with security being the most significant. With the increasing complexity of networks and the diversification of user expectations, it is crucial to prioritize the security and stability of these virtualized systems. Exploring modern literature uncovers a wide range of security techniques and strategies specifically designed for virtualized spectrum infrastructures. An approach that is emphasized involves device authentication. Through the utilization of PBKDF2 and Boolean logic techniques, networks may efficiently authenticate user devices, therefore thwarting any unlawful entry into specific network segments. These authentication procedures serve as the initial barrier, guaranteeing that the network's integrity stays uncompromised.

Nevertheless, despite the use of strong authentication measures, networks remain vulnerable to potential threats. The DDoS assaults have the ability to overwhelm network slices, resulting in a notable decline in the QoS and QoE. Advanced approaches such as the calculation of Renyi entropy are prominently used here. Through the identification and timely resolution of such threats, networks can preserve their peak performance even when confronted with hostile circumstances. The security infrastructure is strengthened by integrating neural networks, particularly the HopFieldNet. These networks optimize the allocation of resources for each network slice by taking into account many criteria, including as signal-to-noise ratio (SINR), throughput, slice capacity, and value. Through the dynamic adjustment of resource distribution in response to real-time demands and conditions, these networks guarantee both efficient and secure utilization of resources.

The multi-access edge computing (MEC) is an additional aspect that improves security and privacy. The MEC effectively mitigates the danger of data breaches by decreasing data transfer across the network and limiting the availability of user information. In addition, the integration of satellite networks offers an extra level of backup, guaranteeing dependable communication in the event of terrestrial network failures. Although these processes provide strong security safeguards, the path to achieving a completely secure virtualized spectrum infrastructure is filled with difficulties. Continuous innovation in security approaches is necessary due to the dynamic nature and intrinsic complexity of these networks. In the future, it is possible that quantum cryptography will be integrated to guarantee secure communication, even against more sophisticated computer threats.

When considering the security aspects of spectrum virtualization, there are several important characteristics and factors that become apparent:

1. **Authentication Mechanisms:** These serve as the foundational layer, ensuring that only authorized devices and users can access the network.
2. **Real-Time Threat Detection and Mitigation:** Methods such as the Renyi entropy calculation provide agile countermeasures against security risks, guaranteeing the integrity of the network.

3. **Dynamic Resource Allocation:** Through the utilization of neural networks and algorithms, systems have the capability to dynamically adjust resource allocation in real time, thereby maximizing both performance and security.
4. **Redundant Communication Pathways:** The integration of satellite networks and other backup transmission techniques guarantees constant connection even in bad situations.

5.5.6 Beyond the Horizon: Future Trends in Spectrum Management and Infrastructure

The field of telecommunications is always changing, fueled by continuous innovation and the unstoppable progress of technology. In this ever-changing environment, spectrum virtualization and network slicing have emerged as powerful and game-changing influences. However, like every emerging technology, the question that emerges is: what does the future hold? By examining recent studies and projecting from present patterns, we can identify many significant future trends that will fundamentally reshape spectrum management and infrastructure.

An important and noticeable trend in literature is the merging of spectrum virtualization with cloud and edge computing [38–40]. When combined, these technologies have the potential to greatly enhance the advantages of one another. Network slicing, supported by spectrum virtualization, enables the provision of personalized and secure network services designed specifically for various applications and situations. Enhancing this adaptability can be achieved by utilizing cloud or edge computing systems, which provide scalability, elasticity, and mobility. This synergy enables the management of network resources, such as spectrum, with unparalleled flexibility and programmability.

The MEC is a prominent feature in this integration. Prior research indicates the possible use of this technology in terrestrial, aerial, and satellite settings [38]. The MEC provides a wide range of advantages, including content caching and computation offloading, efficient exploitation of bandwidth resources, and improved connection dependability. Moreover, its contribution to enhancing the performance of network slicing is remarkable. The MEC facilitates the deployment of cutting-edge network services by minimizing latency and energy usage, particularly for IoT devices. The emergence of edge intelligence and the implementation of VNFs are set to fundamentally transform virtualized spectrum infrastructures through MEC.

Nevertheless, the path toward this unified future is not without of obstacles. To achieve smooth compatibility between spectrum virtualization, cloud, and edge computing, it is essential to have strong frameworks and standards in place. Security, which is always of utmost importance, becomes even more crucial as various technologies merge. Addressing data integrity, user privacy, and system resilience in this intricate setting will require inventive solutions.

By examining the future possibilities, one may imagine a scenario where the combination of spectrum virtualization, cloud computing, and edge computing enables the development of extensive IoT networks, smart cities, and even more realistic augmented and virtual reality experiences. The combination of cloud computing's elasticity, edge computing's real-time processing capabilities, and spectrum virtualization's adaptability has the potential to bring about a new age in telecommunications.

5.6 Conclusion

The field of spectrum management, characterized by the principles of spectrum virtualization and network slicing, is on the verge of a revolutionary period in telecommunications. This chapter explored the complex aspects of this field, emphasizing the fundamental significance of spectrum in influencing the future of wireless communication. As the need for spectrum resources grows rapidly due to the integration of various applications and the merging of voice, data, and multimedia, it becomes essential to develop creative solutions.

During the chapter, we dealt with the numerous difficulties and possibilities that arise from spectrum virtualization. The voyage uncovered the multiple nature of spectrum management, ranging from the technical complexities of scalability and flexibility to the crucial importance of security. The examination of many applications, ranging from smart cities to Industry 4.0, provided a clear illustration of the significant impact that virtualized spectrum infrastructures may have in bringing about transformation. Moreover, the chapter's exploration of future patterns provides light on the course that lies ahead. The combination of spectrum virtualization with emerging technologies such as cloud and edge computing signifies a new age of remarkable efficiency and adaptability in telecommunications.

Essentially, as the telecommunications industry changes, effective spectrum management, supported by virtualization and network slicing, will be crucial in guaranteeing connectivity, stability, and innovation. The trip, although filled with difficulties, is also abundant in potential. To fulfill this pledge, one must be dedicated to ongoing innovation, cooperation, and a forward-looking perspective that goes beyond the current state, paving the way for a unified and interconnected future.

References

1 Neuvo, Y. and Yrjanainen, J. (2002). Wireless meets multimedia-new products and services. *Proceedings of International Conference on Image Processing*, 1. IEEE.

2 Browne, J. (2010). Broadband systems fuel industry growth. *Microwaves & RF* 49 (11): 102–104.

3 Kazovsky, L., Cheng, N., Shaw, W. et al. (2011). *Broadband Access Technologies: An Overview, Broadband Optical Access Networks*, 1–33. Wiley.

4 Faruque, S. (2019). *Radio Frequency Multiple Access Techniques Made Easy*. Cham: Springer.

5 Wang, S. and Ran, C. (2016). Rethinking cellular network planning and optimization. *IEEE Wireless Communications* 23 (2): 118–125.

6 Rappaport, T.S. (2002). *Wireless Communications: Principles and Practice*, 2e. Prentice Hall.

7 Michel, T., Mylène, P., and Emmanuelle, V. (2013). *Wireless Telecommunication Systems*. Wiley.

8 Ayyappan, K. and Kumar, R. (2010). QoS based capacity enhancement for WCDMA network with coding scheme. *International Journal of VLSI Design & Communication Systems* 1 (1): 10–19.

9 Dahlman, E., Parkvall, S., and Sköld, J. (2021). *5G NR The Next Generation Access Technology*. Academic Press.

10 Babaei, A., Park, J., Veque, V., and Trivedi, G. (2010). Dynamic spectrum management in wireless networks. *IEEE Wireless Communications* 17 (4): 4–5.

11 Okon, A.A., Elgendi, I., Sholiyi, O.S. et al. (2020). Blockchain and SDN architecture for spectrum management in cellular networks. *IEEE Access* 8: 94415–94428.

12 Ahmad, W.S.H.M.W., Radzi, N.A.M., Samidi, F.S. et al. (2020). 5G technology: towards dynamic spectrum sharing using cognitive radio networks. *IEEE Access* 8: 14460–14488.

13 Sultana, A., Woungang, I., Zhao, L., and Anpalagan, A. (2019). Two-tier architecture for spectrum auction in SDN-enabled cloud radio access network. *IEEE Transactions on Vehicular Technology* 68 (9): 9191–9204.

14 Walrand, J. and Varaiya, P. (2000). *High-Performance Communication Networks*, 2e. Morgan Kaufmann.

15 Yu, W. and Cfi, J.M. (2002). FDMA capacity of Gaussian multi access channels with ISI. *IEEE Transactions on Communications* 50: 102–111.

16 Cisco Systems, Inc. (2003) Capacity, Coverage, and Deployment Considerations for IEEE 802.11G. White Paper.

17 Gast, M.S. (2013). *802.11ac: A Survival Guide*. O'Reilly Media, Inc.

18 Bektas, C., Bocker, S., Kurtz, F., and Wietfeld, C. (2019). Reliable software-defined RAN network slicing for mission-critical 5G communication networks. *2019 IEEE Globecom Workshops (GC Wkshps)*, 1–6. IEEE.

19 Chowdhury, N.M.K. and Boutaba, R. (2010). A survey of network virtualization. *Computer Networks* 54 (5): 862–876.

20 Carapinha, J. and Jiménez, J. (2009). Network virtualization: a view from the bottom. *Proceedings of the 1st ACM Workshop on Virtualized Infrastructure Systems and Architectures*, 73–80.

21 Shen, X., Gao, J., Wu, W. et al. (2021). Holistic network virtualization and pervasive network intelligence for 6G. *IEEE Communications Surveys & Tutorials* 24 (1): 1–30.

22 Tran, T.X., Hajisami, A., and Pompili, D. (2017). Cooperative hierarchical caching in 5G cloud radio access networks. *IEEE Network* 31 (4): 35–41.

23 Xu, K., Yang, K., and Stojmenovic, I. (2012). Wired and wireless network virtualization. *IEEE Network* 26 (5): 6–7.

24 Ma, S., Wang, B., Zhang, X., and Li, T. (2014). An evolving architecture for network virtualization. *Testbeds and Research Infrastructure: Development of Networks and Communities: 9th International ICST Conference, TridentCom 2014*, Guangzhou, China. (5–7 May 2014), Revised Selected Papers 9, 379–386. Cham: Springer International Publishing.

25 Carapinha, J. and Jiménez, J. (2009). Network virtualization: a view from the bottom. *Proceedings of the 1st ACM Workshop on Virtualized Infrastructure Systems and Architectures*, 73–80. Barcelona, Spain: ACM.

26 Gupta, B.B. and Agrawal, D.P. (2019). *Handbook of Research on Cloud Computing and Big Data Applications in IoT*. IGI Global.

27 Bhatia, S., Motiwala, M., Valancius, V. et al. (2008). Hosting Virtual Networks on Commodity Hardware. Georgia Tech. University. *Tech. Rep.* GT-CS-07-10.

28 Casado, M., Koponen, T., Ramanathan, R., and Shenker, S. (2010). Virtualizing the network forwarding plane. *Proceedings of the Workshop on Programmable Routers for Extensible Services of Tomorrow (PRESTO '10)*, Article no. 8, 1–6. New York, NY, USA: Association for Computing Machinery.

29 Niebert, N., Khayat, I.E., Baucke, S. et al. (2008). Network virtualization: a viable path towards the future internet. *Wireless Personal Communications* 45 (4): 511–520.

30 Barona López, L.I., Valdivieso Caraguay, A.L., García Villalba, L.J., and López, D. (2015). Trends on virtualisation with software defined networking and network function virtualisation. *IET Networks* 4 (5): 255–263.

31 Blenk, A. and Kellerer, W. (2019). Towards virtualization of software-defined networks: a journey in three acts. *IFIP/IEEE Symposium on Integrated Network and Service Management (IM)*, Arlington, VA, USA, 677–682. IEEE.

32 Basta, A., Blenk, A., Dudycz, S. et al. (2018). Efficient loop-free rerouting of multiple SDN flows. *IEEE/ACM Transactions on Networking* 26 (2): 948–961.

33 Blenk, A., Basta, A., Henkel, L. et al. (2018). perfbench: A tool for predictability analysis in multi-tenant software-defined networks. *Proceedings of the ACM SIGCOMM 2018 Conference on Posters and Demos*, Budapest, Hungary, 66–68. ACM.

34 Basta, A., Blenk, A., Lai, Y.-T., and Kellerer, W. (2015). HyperFlex: Demonstrating control-plane isolation for virtual software-defined networks. *2015 IFIP/IEEE International Symposium on Integrated Network Management (IM)*, 1163–1164. IEEE.

35 Blenk, A., Basta, A., and Kellerer, W. (2015). HyperFlex: An SDN virtualization architecture with flexible hypervisor function allocation. *2015 IFIP/IEEE International Symposium on Integrated Network Management (IM)*, 397–405. IEEE.

36 Basta, A., Blenk, A., Hassine, H.B., and Kellerer, W. (2015). Towards a dynamic SDN virtualization layer: control path migration protocol. *2015 11th International Conference on Network and Service Management (CNSM)*, 354–359. IEEE.

37 Giorgetti, A., Scano, D., and Valcarenghi, L. (2021). Guaranteeing slice performance isolation with SDN. *IEEE Communications Letters* 25 (11): 3537–3541.

38 Nadeem, L., Azam, M.A., Amin, Y. et al. (2021). Integration of D2D, network slicing, and MEC in 5G cellular networks: survey and challenges. *IEEE Access* 9: 37590–37612.

39 Patel, W. and Tripathy, P. (2023). eMBB, URLLC, and mMTC 5G wireless network slicing: a communication-theoretic view. *2023 International Conference on Artificial Intelligence and Smart Communication (AISC)*, Greater Noida, India, 1291–1296. IEEE.

40 Chahbar, M., Diaz, G., Dandoush, A. et al. (2021). A comprehensive survey on the E2E 5G network slicing model. *IEEE Transactions on Network and Service Management* 18 (1): 49–62.

41 Subedi, P., Alsadoon, A., Prasad, P.W.C. et al. (2021). Network slicing: a next generation 5G perspective. *EURASIP Journal on Wireless Communications and Networking* 2021 (1): 102.

42 Khan, L.U., Yaqoob, I., Tran, N.H. et al. (2020). Network slicing: recent advances, taxonomy, requirements, and open research challenges. *IEEE Access* 8: 36009–36028.

43 Mekki, M., Arora, S., and Ksentini, A. (2022). A scalable monitoring framework for network slicing in 5G and beyond mobile networks. *IEEE Transactions on Network and Service Management* 19 (1): 413–423.

44 Afolabi, I., Taleb, T., Samdanis, K. et al. (2018). Network slicing and softwarization: a survey on principles, enabling technologies, and solutions. *IEEE Communications Surveys & Tutorials* 20 (3): 2429–2453.

45 Barmpounakis, S., Maroulis, N., Papadakis, M. et al. (2020). Network slicing-enabled RAN management for 5G: cross layer control based on SDN and SDR. *Computer Networks* 166 (C): 106987.

46 Jovanovski, B., Raković, V., and Gavrilovska, L. (2020). Virtualized platforms for network resources orchestration and management. *Journal of Electrical Engineering and Information Technologies* 5 (1), Art. no. 1.

47 Xiao, Y., Hirzallah, M., and Krunz, M. (2018). Distributed resource allocation for network slicing over licensed and unlicensed bands. *IEEE Journal on Selected Areas in Communications* 36 (10): 2260–2274.

48 Li, X., Jiao, K., Jiang, F. et al. (2020). A service-oriented spectrum-aware RAN-slicing trading scheme under spectrum sharing. *IEEE Internet of Things Journal* 7 (11): 11303–11317.

49 Mowla, M.M., Ahmad, I., Habibi, D., and Phung, Q.V. (2017). An energy efficient resource management and planning system for 5G networks. *2017 14th IEEE annual consumer communications & networking conference (CCNC)*, 216–224. IEEE.

50 Alfoudi, A.S.D., Newaz, S.H.S., Otebolaku, A. et al. (2019). An efficient resource management mechanism for network slicing in a LTE network. *IEEE Access* 7: 89441–89457.

51 Vilà, I., Pérez-Romero, J., Sallent, O., and Umbert, A. (2020). Characterization of radio access network slicing scenarios with 5G QoS provisioning. *IEEE Access* 8: 51414–51430.

52 Mei, J., Wang, X., and Zheng, K. (2020). An intelligent self-sustained RAN slicing framework for diverse service provisioning in 5G-beyond and 6G networks. *Intelligent and Converged Networks* 1 (3): 281–294.

53 Mei, J., Wang, X., Zheng, K. et al. (2021). Intelligent radio access network slicing for service provisioning in 6G: a hierarchical deep reinforcement learning approach. *IEEE Transactions on Communications* 69 (9): 6063–6078.

54 Yan, D., Ng, B.K., Ke, W., and Lam, C.-T. (2023). Deep reinforcement learning based resource allocation for network slicing with massive MIMO. *IEEE Access* 11: 75899–75911.

55 Tseliou, G., Samdanis, K., Adelantado, F. et al. (2016). A capacity broker architecture and framework for multi-tenant support in LTE-A networks. *2016 IEEE International Conference on Communications (ICC)*, Kuala Lumpur, Malaysia, 1–6. IEEE.

56 Shen, X., Gao, J., Wu, W. et al. (2020). AI-assisted network-slicing based next-generation wireless networks. *IEEE Open Journal of Vehicular Technology* 1: 45–66.

Section III

6

Spectrum Management for 6G RIS-SWIPT Systems

Neha Sharma[1], Sumit Gautam[1,2], Prabhat Kumar Upadhyay[1], Symeon Chatzinotas[2], and Björn Ottersten[2]

[1] *Department of Electrical Engineering, Indian Institute of Technology Indore, Indore, Madhya Pradesh, India*
[2] *Interdisciplinary Centre for Security, Reliability and Trust (SnT), University of Luxembourg, Esch-sur-Alzette, Luxembourg*

6.1 Introduction

As we look beyond the horizon of the fifth generation (5G) of wireless systems, the dawn of the sixth generation (6G) promises to reshape the way we connect and interact with our digital world. The intensifying use of Internet-of-Things (IoT) devices, having applications from augmented reality and autonomous vehicles to remote surgery and smart cities, has underscored the necessity of addressing their energy needs [1]. Batteries are like the lifeblood of these devices, which however often come with certain limitations that constrain their deployment in remote or inaccessible areas. Thus, it becomes almost essential to develop a system that addresses the data-intensive applications and the efficient energy utilization.

Spectrum management is a critical endeavor in modern wireless communication systems to ensure efficient utilization of the limited frequency bands available. One innovative approach for enhancing spectrum optimization is through the integration of reconfigurable intelligent surfaces (RIS) [2]. These intelligent surfaces, capable of altering and manipulating radio waves, hold the potential to revolutionize the wireless networks. By intelligently adjusting the phase and amplitude of reflected signals, RIS can steer electromagnetic waves toward desired direction with enhanced signal strength [3]. This manipulation in terms of each element's phase shift, reflectivity, directivity, etc., is facilitated by an external stimulus circuit, which includes components like PIN diodes, FPGAs, microcontrollers, and more [4]. This technology not only aids in augmenting

Intelligent Spectrum Management: Towards 6G, First Edition.
Edited by Sridhar Iyer, Anshuman Kalla, Onel Alcaraz López, and Chamitha De Alwis.

coverage and capacity but also contributes to energy efficiency by reducing the need for high-power transmissions. The multi-element RIS technology, similar to programmable mirrors, is a passive device (only reflection, no amplification), made up of several tiny meta-materials. It acts as a mediator between transmitter and receiver by adapting to the signal path and provides swift connectivity to remote areas where direct communication is being hindered by physical barriers. The authors in [5] explore the support of spectrum management by employing directional beamforming of RIS to alleviate internode interference within an IoT network. The spectrum-sharing strategy using RIS for successive interference cancellation (SIC) is given by Jiang and Yu [6], in which numerous pairs of transceivers engage in simultaneous communication over a shared spectrum. Exploiting the deep learning (DL) framework for performance prediction, particularly in situations where SIC is imperfect within relay networks (as discussed in [7]), can also find analogous applications in the context of RIS, as indicated in [8]. The spectrum management using RIS also finds its application in cognitive radio networks [9], vehicular technology [10], spectrum coexistence [11], edge networks [12], etc. Hence, RIS presents a promising avenue for transforming the way we manage and utilize the electromagnetic spectrum, fostering a new era of smart and adaptable wireless communication systems.

Equally important, the energy harvesting (EH) is an eco-friendly transformative solution for enabling IoT devices to draw power from electromagnetic signal, thereby extending their operational life and reducing reliance on conventional power sources. The corresponding technique is termed as simultaneous wireless information and power transfer (SWIPT). It has two widely recognized protocols, known as time-switching (TS) and power-splitting (PS) [13]. The TS approach involves partitioning the overall time duration into segments for implementing the SWIPT mechanism, whereas the PS method entails division of overall power into portions for SWIPT throughout the complete time frame. In an effort to quantify the EH metric at a user, numerous studies in the literature have assumed linear EH models, such as piecewise linear EH and the constant-linear-constant (CLC) EH model [14]. However, the linear EH model falls short in accurately representing real-world scenarios and may lack practical implementation. Consequently, a non-linear EH (NL-EH) model based on sigmoidal/logistic functions was introduced in [15].

Furthermore, the transformational synergy comprising of RIS with SWIPT technology presents opportunities for enhanced data transmission rates, extended coverage, and improved energy efficiency [16]. Table 6.1 bifurcates the recent study on RIS-SWIPT systems present in literature on the basis of EH-model used, optimization or performance analysis carried out, SWIPT architecture, and single or multiple RIS scenario. The significant contribution acknowledging above points is given in Section 6.1.1.

Table 6.1 Recent study on RIS-SWIPT.

References	EH model	Objective function	SWIPT architecture	RIS
Sharma et al. [17]	CLC	a)	Hybrid TS-PS	Single
Dhok et al. [18]	CLC	a)	Hybrid TS-PS	Single
Do et al. [19]	—	a)	—	Multi
Galappaththige et al. [20]	—	a)	—	Multi
Yildirim et al. [21]	—	a)	—	Multi
Aldababsa et al. [22]	—	No. of elements	—	Multi
Sharma and Gautam [23]	NL-EH	EH	TS	Single
Sharma and Gautam [24]	NL-EH	Rate	PS	Single
Sharma [25]	NL-EH	Rate, EH and power	PS	Single
Sharma and Gautam [26]	NL-EH	Rate, EH and power	TS	Single
Wang et al. [27]	NL-EH	Sum rate	—	Single
Xu et al. [28]	NL-EH	Total energy	PS	Single
Mohamed et al. [29]	Linear	Rate, EH	PS	Single
Nwufo et al. [30]	Linear	Energy efficiency	Separated	Multi
Galappaththige and Baduge [31]	NL-EH	a)	TS/PS	Multi
Le and Le [32]	Linear	a)	only EH	Multi
Mohamed and Aïssa [33]	Linear	a)	only EH	Multi

a) Performance analysis.

6.1.1 Motivation

The aspects of spectrum management using multiple RIS-assisted wireless communication surfaces is the aim of this study. Various linear EH model variants (such as linear, piecewise, or CLC models) are compared with NL-EH model highlighting NL behavior of energy conversion efficiency. Since there are fewer studies considering multi-RIS scenarios employing SWIPT, thus to explain flexibility of placement and arrangement in multi-RIS assisted settings is considered, which tries to fill the gap in literature.

Furthermore, a case study of optimizing several objectives comprising the SWIPT mechanism at the IoT user equipment (U), subjected to certain Quality-of-Service (QoS) constraints for improved performance, is given. It utilizes the divide-and-conquer methodology, for an iterative algorithm to optimize the PS ratio and transmit power, aiming to boost the rate. Few aspects from the literature will be examined by the numeric interpretations from the study.

The chapter is organized in following way: Section 6.2 deals with the EH models, Section 6.3 with multiple RIS scenarios comprising of RIS selection strategies and SWIPT protocol. It is followed by Section 6.4 stating the case study with impact of RIS arrangements, Section 6.5 with spectrum management, and recent advancements are provided in Section 6.6. The chapter is summarized with conclusion in Section 6.7.

6.2 Energy Harvesting Models

After several decades of research in the field of wireless radio-frequency (RF) power transfer, it is evident that the backbone of all EH circuits is the rectenna, composed of an antenna and a rectifier. The standard approach involves converting the incoming wireless RF signal into a DC equivalent, using single or multiple diodes. These series of diodes introduce nonlinearity to the system operation. It is crucial to closely or exactly model the nonlinearity in the EH circuit, especially concerning SWIPT. The literature is served with various types of linear and approximated nonlinear-type models listed below and plotted in Figure 6.1 [14, 34].

6.2.1 Linear EH Model

The linear EH model stands out as the most basic and commonly employed in a significant portion of literature related to SWIPT. It expresses the harvested power

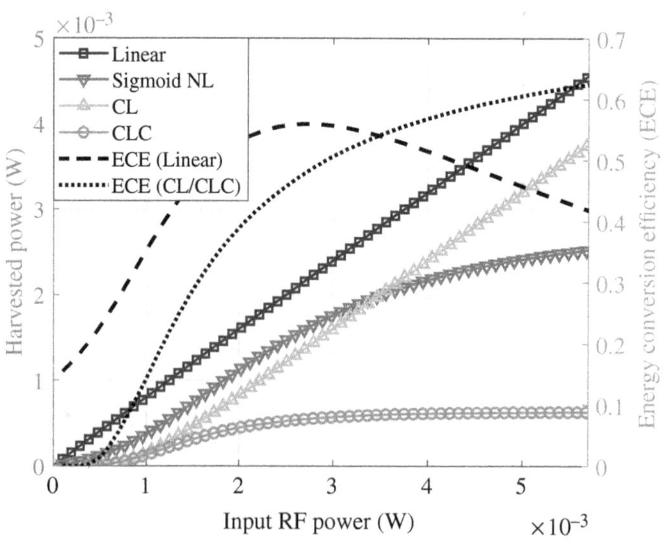

Figure 6.1 Linear and nonlinear EH with energy conversion efficiency.

as a linear function of input RF power (P_{in}) in the following manner:

$$P_{lin} = \zeta_{lin} P_{in}, \tag{6.1}$$

where $\zeta_{lin} \in (0, 1]$ is the constant energy conversion efficiency (ECE) of the linear EH receiver. The nonpracticality of linear model is acknowledged by subsequent models.

6.2.2 Constant-Linear EH Model

The constant-linear (CL) model considers the sensitivity of the receiver (P_{sen}), a threshold below which RF conversion does not occur, given as:

$$P_{CL} = \begin{cases} 0, & P_{in} \in [0, P_{sen}], \\ \zeta_{CL}(P_{in} - P_{sen}), & P_{in} \in [P_{sen}, \infty), \end{cases}$$

where $\zeta_{CL} \in (0, 1]$ is the fixed ECE of the CL-EH receiver. This model lacks in considering the input power saturation.

6.2.3 Constant-Linear-Constant EH Model

The constant-linear-constant (CLC) model considers the sensitivity of the receiver (P_{sen}) as well as saturation power level (P_{sat}) given as:

$$P_{CLC} = \begin{cases} 0, & P_{in} \in [0, P_{sen}], \\ \zeta_{CLC}(P_{in} - P_{sen}), & P_{in} \in [P_{sen}, P_{sat}], \\ \zeta_{CLC}(P_{sat} - P_{sen}), & P_{in} \in [P_{sat}, \infty). \end{cases}$$

In between $P_{sen} - P_{sat}$ levels, linear operation of RF conversion is carried with $\zeta_{CLC} \in (0, 1]$ efficiency.

6.2.4 Non-linear EH Model

The sigmoid/logistic function based NL-EH model truly replicates NL behavior of diodes; it is given in [15] as:

$$P_{NL} = \frac{E'}{1 - \Phi}\left(\frac{1}{1 + exp(-aP_{in} + ab)} - \Phi\right), \tag{6.2}$$

where $\Phi = (1 + exp(ab))^{-1}$ guarantees a zero input for a zero output in the model. E' is the circuit's saturation point, a and b are related to the circuit's capacitor and diode turn-on voltage, respectively. A curve-fitting tool is used to analyze the experimental data and to find the values of the constant parameters (E', a, b) [15, 35].

Note: A different nomenclature is adopted by some authors while mentioning CLC model as NL or piecewise NL model [17–19, 36, 37]. However, we shall restrict our discussion to sigmoidal function-based NL-EH model in this chapter.

Figure 6.1 sketches the NL nature of ECE for above-mentioned models $\zeta_k \forall k \in [lin, CL, CLC]$. ECE is considered to be a constant value for linear, CL and CLC models, which is not a true case due to circuit nonlinearity [38].

6.3 Multiple RIS Scenario

We consider a multiple RIS-assisted communication system, comprising single source base station (BS), single IoT user (U) node, and N different RISs placed randomly apart from each other. The nth RIS $\{R_n\}_{n=1}^{N}$, have M_n total reflecting elements. There may be possibility for varying RISs sizes, whose effect will be studied in Section 6.4.3. Both the BS and U are equipped with single antenna, making it a half-duplex (HD) system. It is noteworthy that there is no possible direct communication between BS and U, due to the presence of physical obstructions. The IoT node is capable of SWIPT operation, that is, EH as well as information decoding (ID) simultaneously. The communication is from the BS to the ith element of nth RIS (R_n), where $i \in [1, M_n]$, then directed to U. The depiction of communication path for the first RIS is given in Figure 6.2. A low-mobility environment is considered with deployment of RISs on building walls and billboards, leading to

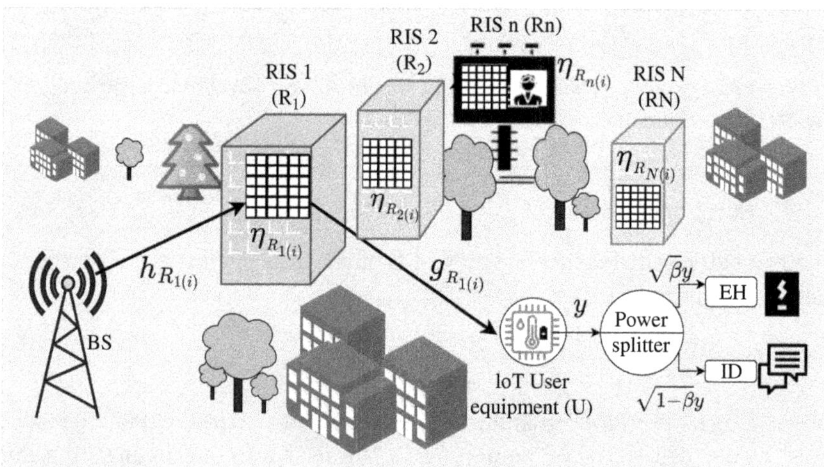

Figure 6.2 System model for multi-RIS assisted IoT system employing PS-SWIPT on the received signal ($y \in [\hat{y}, \bar{y}]$).

the absence of interference between them. Thus, the reflection from RISs will not interfere with each other upon reaching U [19].

The BS-$R_{n(i)}$ (BS to ith element of R_n RIS) link follows complex normal distribution, given as $\tilde{h}_{R_{n(i)}} \sim \mathcal{CN}(0, \sigma^2_{h_{R_{n(i)}}})$. Similarly, $R_{n(i)}$-U link's complex channel distribution is $\tilde{g}_{R_{n(i)}} \sim \mathcal{CN}(0, \sigma^2_{g_{R_{n(i)}}})$. $\tilde{h}_{R_{n(i)}} = h_{R_{n(i)}} e^{j\phi_{R_{n(i)}}}$ and $\tilde{g}_{R_{n(i)}} = g_{R_{n(i)}} e^{j\psi_{R_{n(i)}}}$ are the respective channel coefficient given in polar form, having magnitudes ($h_{R_{n(i)}} = |\tilde{h}_{R_{n(i)}}|, g_{R_{n(i)}} = |\tilde{g}_{R_{n(i)}}|$) following Rayleigh distribution and phase as $\phi_{R_{n(i)}} \in [-\pi, \pi]$ and $\psi_{R_{n(i)}} \in [-\pi, \pi]$ following uniform distribution. The phase shift matrix associated with the nth RIS (R_n) is $\Theta_n = diag([\eta_{R_{n(1)}} e^{j\theta_{R_{n(1)}}}, \eta_{R_{n(2)}} e^{j\theta_{R_{n(2)}}}, \dots, \eta_{R_{n(M_n)}} e^{j\theta_{R_{n(M_n)}}}])$. The amplitude of reflection coefficient is $\{\eta_{R_{n(i)}}\}_{i=1}^{M_n} \in (0, 1]$ and $\theta_{R_{n(i)}} \in [-\pi, \pi]$ is phase shift of the ith elements ($i \in [1, M_n]$), responsible for incoming signal steering toward the user. Whether it is viable to use all N RISs or the best one, will be decided as per the strategies proposed in Section 6.3.1.

6.3.1 Multi-RIS Selection Strategies

When several RISs are in the vicinity, their selection becomes important for optimized performance. There are two broad approaches to RIS selection mentioned in the literature, namely exhaustive RIS approach (ERA) and optimum RIS approach (ORA) [19], respectively.

6.3.1.1 Exhaustive RIS Approach (ERA)

In this approach, all the RISs with all its elements $\left\{\{R_{n(i)}\}_{i=1}^{M_n}\right\}_{n=1}^{N}$, assist the signal transmission (s) between BS and U at same time-frequency scale. Each RIS aims to maximize the SNR at U, thus intelligently steering the incoming signal (multiple copies from BS) at U, with the assumption of no interference among them as in [19, 20]. P is BS transmission power and additive white Gaussian noise (AWGN) has zero mean and N_0^2 variance. The received SNR at U following the path, BS $\rightarrow \left\{\{R_{n(i)}\}_{i=1}^{M_n}\right\}_{n=1}^{N} \rightarrow$ U by employing channel estimation techniques for distributed RIS as reported in [39], resorts to an assumption of perfect channel state information (CSI) at all RISs. The maximized SNR is

$$\hat{\gamma} = \frac{P}{N_0^2}\left|\sum_{n=1}^{N}\sum_{i=1}^{M_n} g_{R_{n(i)}} \eta_{R_{n(i)}} h_{R_{n(i)}}\right|^2 = \frac{P\hat{\mathbb{Y}}}{N_0^2}, \tag{6.3}$$

6.3.1.2 Optimum RIS Approach (ORA)

The ORA approach uses only the most suitable RIS for the signal s transmission, instead of all the $\left\{\{R_{n(i)}\}_{i=1}^{M_n}\right\}_{n=1}^{N}$ RISs. This way, one can save energy, resources, and reduce complexity. The ORA scheme also makes the receiver's job easier, as it has to drive only single R_n RIS's M_n reflecting signals, unlike the ERA scheme. The 'suitability' of nth RIS out of N RISs, lies in its ability to provide the highest SNR. This is carried out by first calculating the received signal via individual R_n RIS's reflection, followed by finding SNR. The maximized SNR provided by the ith element of R_n by nullifying the channel phases, same as ERA (provided perfect CSI is known), is represented by

$$\gamma_n = \frac{P}{N_0^2}\left|\sum_{i=1}^{M_n} g_{R_{n(i)}} \eta_{R_{n(i)}} h_{R_{n(i)}}\right|^2 = \frac{P\mathbb{Y}_n}{N_0^2}. \tag{6.4}$$

The RIS selection is mathematically specified as [19, 21, 22], expressing SNR for ORA approach as follows:

$$\overline{\gamma} = \max_{n=1,\ldots,N} \gamma_n = \frac{P\overline{\mathbb{Y}}}{N_0^2}. \tag{6.5}$$

The $n^* = arg(\overline{\gamma})$ provides the 'suitable' (n^*) RIS having maximum SNR. In (6.5), $\overline{\mathbb{Y}} = \max_{n=1,\ldots,N} \mathbb{Y}_n$, implies that the highest SNR selection depends on the channel conditions and number of elements of RIS.

Remark *In further discussion, the SNR is represented as $\gamma \in [\widehat{\gamma}, \overline{\gamma}]$ adopting either of ERA (6.3) or ORA (6.5) approach, respectively, according to the choice of operation.*

6.3.2 SWIPT Protocol

In this section, we consider PS-SWIPT protocol to be followed at U, due to its spectral efficient and resource-saving advantages, resulting in dividing the received signal y into two fractions. A part $\sqrt{\beta}y$ is used for EH, and remaining part for ID $\sqrt{1-\beta}y$. β represents PS factor. The EH part of the PS protocol at U involves NL-EH, which is a modified version of (6.2). The corresponding expression is defined as [38]:

$$\widehat{E_h} = \mathcal{E}\left(\frac{1}{1+exp(-\beta a P\widehat{\mathbb{Y}} + ab)} - \Phi\right),$$
$$\overline{E_h} = \mathcal{E}\left(\frac{1}{1+exp(-\beta a P\overline{\mathbb{Y}} + ab)} - \Phi\right), \tag{6.6}$$

where $\mathcal{E} = \frac{E'}{1-\Phi}$ and $\widehat{E_h}, \overline{E_h}$ indicating ERA or ORA approach respectively, adopted at a time. The ID part of the PS protocol at U involves throughput/data rate calculations by

$$\widehat{R_u} = \log_2\left(1 + \frac{(1-\beta)P\widehat{\mathbb{Y}}}{N_0^2}\right),\ \overline{R_u} = \log_2\left(1 + \frac{(1-\beta)P\overline{\mathbb{Y}}}{N_0^2}\right). \tag{6.7}$$

Here $\widehat{R_u}, \overline{R_u}$ are rate for ERA and ORA approach, respectively, modifying their respective SNR as given in Eqs. (6.3), (6.5). We outline the subsequent triplet (rate, EH and channel parameter) to enhance the clarity of the notation as:

$$(R_u, E_h, \mathbb{Y}) := (\widehat{R_u}, \widehat{E_h}, \widehat{\mathbb{Y}}) : \text{ For ERA; } (\overline{R_u}, \overline{E_h}, \overline{\mathbb{Y}}) : \text{ For ORA.} \tag{6.8}$$

In Section 6.4, we take case study of rate maximization along with its proposed solution.

6.4 Case Study

This section presents the case study for the maximization problem related to multiple RIS scenarios and its solution using a heuristic algorithm.

6.4.1 Rate Maximization

This subsection presents the problem formulation associated with the maximization of the data rate. We aim to determine the optimal values of the PS ratio and transmit power, while considering constraints on rate, EH, transmit power, and PS factor. The $R_u \in [\widehat{R_u}, \overline{R_u}]$ denotes rate for ERA and ORA approach respectively [refer (6.8)]. The objective of problem ($P1$) is to maximize the rate for adopted ERA/ORA. Mathematically, the overall optimization problem is defined as follows, where β and P are the relevant parameters

$$\text{(P1): } \underset{\beta,P}{\text{Maximize}} \quad R_u(\beta, P) \tag{6.9}$$

$$\text{Subject to:} \quad R_u \geq R_{th}, \tag{6.10}$$

$$E_h \geq \xi_{th}, \tag{6.11}$$

$$P \leq P_{\max}, \; 0 \leq \beta \leq 1. \tag{6.12}$$

Here R_{th} and ξ_{th} denote the demanded rate and EH, respectively. The transmit power is bounded by $P_{\max}$. The difficulty of solving problem ($P1$) originates from its nonconvex and coupled-integer nature, where both β and P affect the objective and the constraints. To tackle ($P1$), we propose Algorithm 6.1, which is motivated by an iterative divide-and-conquer approach.

Algorithm 6.1 is an iterative process that begins by initializing the values of $R_u(\beta, P)$. These values are within a feasible range to solve the problems ($P1$). The algorithm then finds the optimal values for β^*, P^*, and recalculates $R_u(\beta, P)$. The iteration continues as long as the present value exceeds the previous value, until a lower value is obtained, or iteration limit j' is touched. This process can be summarized as follows:

The above algorithm inherently provides a suboptimal solution to problem ($P1$).

Algorithm 6.1 Iterative algorithm for (P1) maximization

Initialize $R_u(\beta, P)$ with a feasible value and $j = 2$.
REPEAT
Solve $(P1)$ by fixing P and find β^*.
Solve $(P1)$ with $\beta = \beta^*$ and find P^*.
Compute $R_u(\beta, P)[j+1]$ with β^* and P^* and save for next comparison.
IF $\left(R_u(\beta, P)[j+1] < R_u(\beta, P)[j]\right)$
STOP
ELSE return $P = P^*, j = j+1$, and store $R_u(\beta, P)[j+1]$.
UNTIL $R_u(\beta, P)[j] \geq R_u(\beta, P)[j-1]$ is true & $j \leq j'$

6.4.2 Simulation Setup

In this section, we evaluate the efficacy of the proposed solutions for the given problem through simulations. The suggested iterative algorithm is employed to maximize the rate and EH for ERA/ORA, utilizing MATLAB's non-convex program solver *fmincon* $(\cdot)$. The ITU-R short-range outdoor site-general model is employed to generate channels. This model includes the path-loss (in dB), which is represented by ITUR P Series [40]. The fixed coefficients λ, α, σ, and Γ are considered for general urban high-rise/low-rise/suburban environment given by Series [40], while the transmitter–receiver separation distance d (5–660 m). f is the operating frequency in GHz range (0.8–73 GHz) [25]. The specific values of common parameters are given in Table 6.2, unless otherwise stated.

6.4.3 Impact of Various RIS Arrangements

The numeric interpretations from the case study will help us to answer the questions present in literature about the RIS elements, RIS surfaces, their placements, either to choose all elements or smartly place them, etc. will be

Table 6.2 Parameters used in simulations.

Parameters	Values	Parameters	Values
λ	2.12	R_{th}	10 bpcu
α	29.2	P_{max}	10 W
Γ	2.11	N_s	1000
σ	5.06	E'	2.8 mJ
N_0^2	−120 dBm	a	1500
T_s	1 ms	b	0.0022

Table 6.3 Emulation environment for five scenarios.[a)]

Scenario No.	RIS surfaces	RIS elements	Distance (m)
1	1	$R_1 = 150$	$d_{bR} = d_{Ru} = 8$
2	2 (series)	$R_1 = 75$	$d_{bR_1} = 6, d_{R_1u} = 10$
		$R_2 = 75$	$d_{bR_2} = 10, d_{R_2u} = 6$
3	2 (parallel)	$R_1 = R_2 = 75$	$d_{bR_1} = d_{R_1u} = d_{bR_2} = d_{R_2u} = 8$
4	3 (series)	$R_1 = 50$	$d_{bR_1} = 5, d_{R_1u} = 11$
		$R_2 = 50$	$d_{bR_2} = 8, d_{R_2u} = 8$
		$R_3 = 50$	$d_{bR_3} = 11, d_{R_3u} = 5$
5	3 (parallel)	$R_1 = R_2 = 50$	$d_{bR_1} = d_{R_1u} = d_{bR_2} = d_{R_2u} = 8$
		$R_3 = 50$	$d_{bR_3} = d_{R_3u} = 8$

a) Source: Aldababsa et al. [22]/IEEE.

answered in this section. To obtain insights about the optimized values, combined effect of number of RIS surfaces, individual RIS size, multiple RIS arrangement, and $BS - R_n/R_n - U$ distance, five essential scenarios are considered (detailed in Table 6.3). The sketch of the scenarios is inspired from ([22], Figure 3). The total distance between BS and U is 16 m and total RIS elements are 150 (distributed/nondistributed). The first scenario describes a single (nondistributed) RIS surface usage with 150 elements. The second and third scenarios give distributed series placement of two and three RIS surfaces respectively, each with 75 and 50 elements. RIS surfaces are placed parallel-wise in third and fifth scenarios with similar number of surfaces and element number as that of second and fourth case, respectively.

6.4.3.1 Distributed or Collective RIS Elements: Which One to Choose?

For Figure 6.3, the operating frequency of 0.915 GHz is used as it is experimentally concluded for more efficient EH charging [41].

In Figure 6.3a, ERA and ORA approaches are compared for the considered five scenarios. The first scenario (single RIS) outperforms other scenarios as it has the highest number of RIS elements. ERA approach gives higher rate for all scenarios than ORA as all RIS surfaces with all its elements are active giving higher SNR. In addition, series placement performs better than parallel due to lesser $BS - R_n/R_n - U$ distance. Thus, collective placement of elements on one RIS is best choice than distributed RIS surfaces considering the single user scenario.

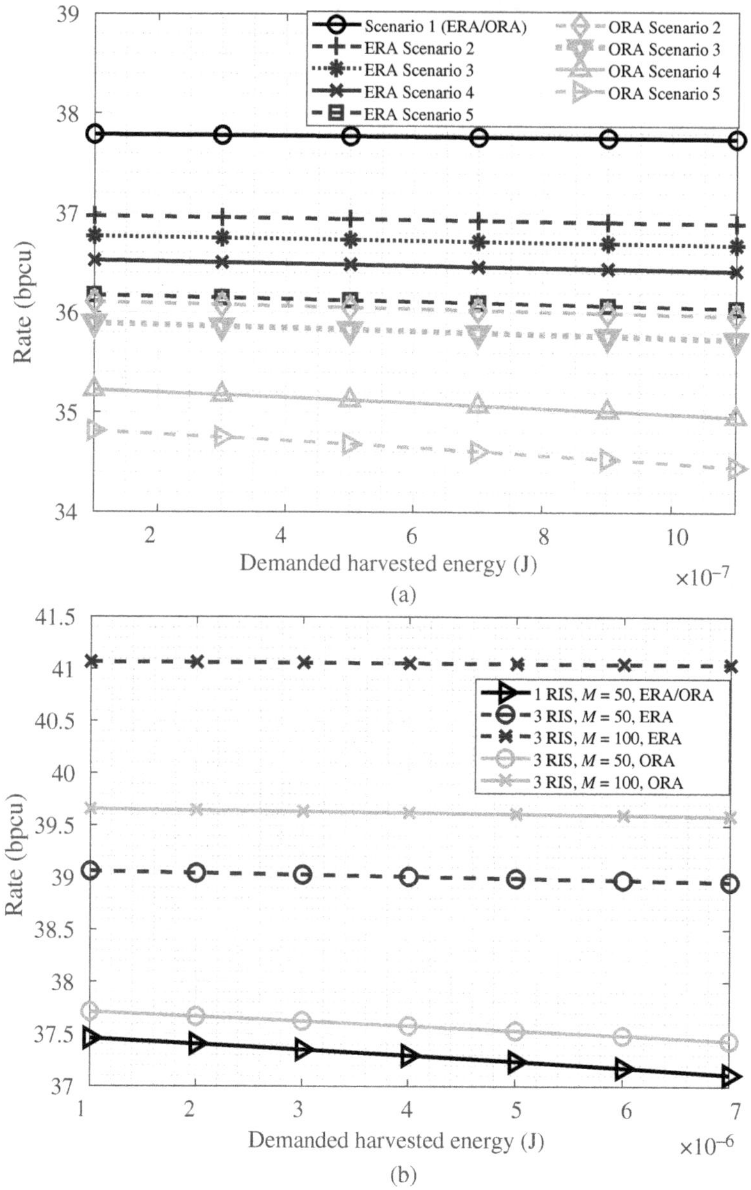

Figure 6.3 Comparative study of ERA and ORA for (a) five scenarios. Source: Aldababsa et al. [22]/IEEE (b) RIS size and number of RIS surfaces.

6.4.3.2 Comparing ERA and ORA Strategies

The emulation of Figure 6.3b considers ERA/ORA case for equal $BS - R_n/R_n - BS$ distance placement (parallel for multiple case). It compares the effects of using single and multi-RIS, provided all RIS surfaces having the same elements. It is clearly observed that by increasing RIS surfaces (single to three), rate increases for both approaches (ERA better than ORA). Thus, rate increases with RIS elements, RIS surfaces, lower demanded harvested energy and ERA approach. This plot gives essential inference that for extended multiuser setting, utilizing multi-RIS does not degrade system performance.

6.4.3.3 Through the Lens of Outage Probability

Analyzing the outage probability with transmit SNR for ORA scenario (saving resources), author in [22] suggested that Scenario 1 surpasses the other four scenarios in system coding gain. This is due to the fact that, with RIS selection, only the reflecting elements of the chosen RIS are active at any moment, while in a single RIS setup, all reflecting elements are active simultaneously. Concentrating all reflecting elements into a single RIS generally yields better outcomes than spreading them across multiple RISs, even if all RISs are active. Beamforming can be utilized for enhancing multi-RIS performance compared to a single RIS. In addition, increasing the number of RISs degrades performance. Comparing the scenario where RISs are sequentially placed between the source and destination to the scenario where they are arranged in parallel, Scenarios 2 and 4 perform better than Scenarios 3 and 5 for both two and three RISs. This is due to the fact that the distance between communicating agencies (BS-RIS/RIS-U) directly influences the system's path-loss, thereby impacting the coding gain.

6.5 Spectrum Management

Spectrum management can be assured with the mere reflection property of RIS giving higher spectral efficiency. It does not require special tuning for a particular frequency. In Figure 6.4, we have taken four frequencies corresponding to four different technologies/protocols; 0.865 GHz (LoRa in India), 0.915 GHz (Zigbee in US), 2.4 GHz (WiFi IEEE 802.11), and 2.48 GHz (Bluetooth). Analysis of spectral efficiency and EH is performed using single RIS with 100 elements. It is evident from Figure 6.4 that spectral efficiency and EH are more at lower frequencies and lesser BS-RIS/RIS-U ($d_{br} = d_{ru} = d$) separation distance. With the assistance of RIS, the IoT node will be able to harvest as high as approximately 0.4 mJ of energy for LoRa protocol (0.865 GHz), additionally providing spectral efficiency/data rate of 40 bpcu, at $d = 5$ meters of BS-RIS/RIS-U separation distance. The observed EH levels and data rates prove adequate for energizing IoT equipment positioned

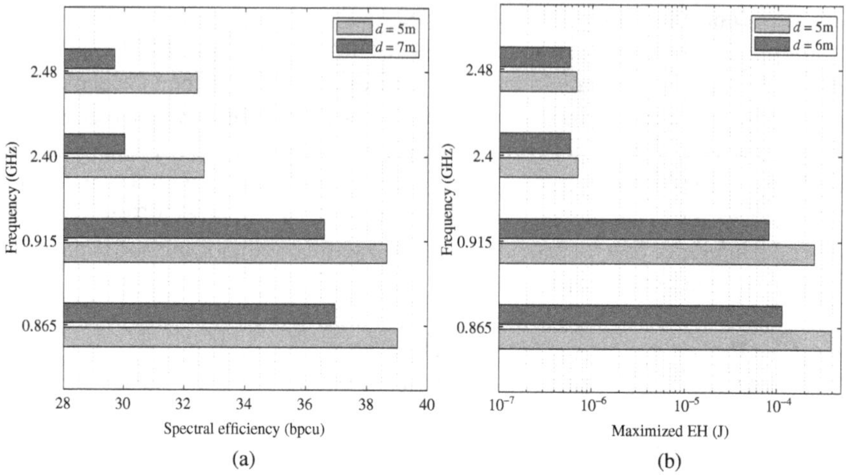

Figure 6.4 Comparative frequency analysis of various protocols (spectrum management) with varying distance in terms of (a) spectral efficiency (bits per channel use) and (b) EH.

in remote locations providing uninterruptible connectivity. The basic conclusions drawn in Section 6.4.3 for ERA/ORA approach, applies in same vein to spectrum management as well. Addressing the complex issue of achieving massive connectivity for IoT devices while ensuring ultra-reliable low-latency communications (URLLC) necessitates spectrum partitioning and scheduling optimization. An algorithm based on data-driven genetic spectrum partitioning (DDGSP) holds promise in this regard. The idea of incorporating intelligent spectrum learning (ISL) functionalities onto RIS in order to predict or estimate the interfering devices based on incoming signals for scenarios involving multiple users, can be worked out with convolution neural networks (CNN) model. However, all the mentioned works are beyond the scope of this chapter and are open for further investigation. As the study considers an optimization problem formulation, including comparative results. It will provide a new dimension to the results from the proposed technique.

6.6 Recent Advancements

6.6.1 Beyond Diagonal RIS (BD-RIS) Systems

Extensive research focus has been directed toward the utilization of traditional RIS employing diagonal phase shift matrices, such that there is no interconnection between the RIS element and they are independently connected to its ground load. The conventional RIS architecture restricts its ability to manipulate passive beamforming, thus researchers are exploring BD-RIS [42]. It offers greater flexibility in manipulating both diagonal and nondiagonal elements of the scattering matrix, significantly enhancing performance in various wireless systems. Consequently, while considering RIS having discrete values, BD-RIS gives increased proficiency. The conventional RIS covers only the half-space, the BD-RIS utilizing appropriate group-connected reconfigurable impedance network and antenna array arrangement can support the hybrid and multisector modes to realize full-space coverage. It has advantages for simplifying practical deployments and enhancing spectral efficiency. After counting for various advantages of BD-RIS, it has greater issues from the viewpoints of equipment deployment, structural planning, RF deficiencies, signal analysis, and wideband modeling.

6.6.2 RIS-Assisted Free Space Optics (FSO) Communication

Free space optics (FSO) communication provides several benefits, such as increased bandwidth and higher throughput, compared to RF communication. As licensed spectrum is limited and expensive, FSO is a better candidate for unlicensed spectrum management. However, FSO links are negatively impacted by atmospheric conditions like fog, cloud or smog, turbulence, and pointing errors and also by obstacles between the transmitter and receiver. FSO when paired with RIS potentially enhances its reliability and coverage, specially overcomes the barriers in the Line-of-Sight (LoS) path [43].

6.6.3 RIS-Assisted Vehicular-to-Everything (V2X) Communication

The Internet-of-Vehicles (IoV) plays a crucial role in intelligent transportation systems (ITS) and is advancing swiftly through the integration of smartly connected vehicles. IoV enables vehicles to interact via V2X communications. Two primary technologies have been employed for V2X communications, first one is dedicated short-range communication (DSRC), and another is vehicular networks based on cellular technology. DSRC is popular commercially but fails for dense and high-mobility environments, adversely affecting coverage, QoS and data rate, leading to unreliable channel delays. Utilizing RIS for V2X scenarios has potential for improving the QoS [44]. Though modeling RIS-V2X scenarios with practical delay-Doppler shift is challenging, opening the doors for new the research area.

6.6.4 RIS-Aided Integrated Sensing and Communications (ISAC)

The convergence of wireless communication networks and sensing systems has been under investigation for several years. Initially, efforts have centered on enabling these systems to coexist on the same frequency spectrum, rather than achieving a full integration on a shared platform. However, ISAC's vision for 6G aims to tightly integrate these functionalities. This involves using a single hardware platform that transmits a unified waveform capable of carrying information symbols and performing environmental sensing simultaneously [45]. This could be accomplished by implementing extensive multiple antenna arrays, the initial answer is to deploy massive multiple-input multiple-output (MIMO), but they poses problem of heavy power consumption. The energy-efficient solution is to use wireless transceivers with reconfigurable metasurfaces, enabling efficient transmit and/or receive beamforming. Having numerous adjustable parameters for signal optimization seems essential for effectively integrating communication and sensing, enhancing spectral efficiency. The key areas of research in RIS-aided ISAC involve localization, multitarget detection, improving multiuser wireless communication, and joint design of transmit beamformers and RIS matrix for interference mitigation.

6.7 Conclusion

This chapter investigates the spectrum management using multiple RIS-assisted PS-SWIPT system employing NL-EH at the IoT user equipment assuming (ni.i.d) RIS channels. The outcomes revealed that allocating an equal number of reflecting elements across multiple RISs leads to reduced system performance compared to concentrating them all on a single RIS. Additionally, this study explored flexible scenarios concerning RIS placement in relation to the BS and U, and their influence on the overall system performance was thoroughly discussed. Lastly, spectrum management for several standards/protocols was studied, claiming RIS to be more spectral efficient and harvesting higher energy at lower frequencies. Incorporating the hybrid TS-PS protocol into the system enhances its potential, allowing EH at the RIS elements within multi-user and multi-RIS scenarios. RIS can be merged with existing well-developed relay technology full-filling back-end processing. It also finds applications in unmanned aerial vehicle (UAV), V2X, ISAC, FSO, and high-altitude platform station (HAPS) for improved performance, ensuring swift communication specially during emergency situations. A new area of BD-RIS is introduced for complete utilization of RIS phase shift matrix and not just the diagonal one. Future investigations should aim to surpass the idealized assumptions in RIS analysis related to perfect CSI and hardware impairments.

References

1 Ding, J., Nemati, M., Ranaweera, C. and Choi, J. (2020). IoT connectivity technologies and applications: a survey. *IEEE Access* 8: 67646–67673. https://doi.org/10.1109/ACCESS.2020.2985932.

2 Di Renzo, M. and Aravanis, A.I. (2021). Catching the 6G wave by using metamaterials: a reconfigurable intelligent surface paradigm. In: *Shaping Future 6G Networks: Needs, Impacts, and Technologies*, 69–87. Wiley.

3 Di Renzo, M., Zappone, A., Debbah, M. et al. (2020). Smart radio environments empowered by reconfigurable intelligent surfaces: how it works, state of research, and the road ahead. *IEEE Journal on Selected Areas in Communications* 38 (11): 2450–2525. https://doi.org/10.1109/JSAC.2020.3007211.

4 Björnson, E., Özdogan, Ö., and Larsson, E.G. (2020). Reconfigurable intelligent surfaces: three myths and two critical questions. *IEEE Communications Magazine* 58 (12): 90–96. https://doi.org/10.1109/MCOM.001.2000407.

5 Mondal, A., Al Junaedi, A.M., Singh, K., and Biswas, S. (2022). Spectrum and energy-efficiency maximization in RIS-Aided IoT networks. *IEEE Access* 10: 103538–103551. https://doi.org/10.1109/ACCESS.2022.3209823.

6 Jiang, T. and Yu, W. (2022). Interference nulling using reconfigurable intelligent surface. *IEEE Journal on Selected Areas in Communications* 40 (5): 1392–1406. https://doi.org/10.1109/JSAC.2022.3143220.

7 Shukla, A.K., Yadav, K., Upadhyay, P.K., and Moualeu, J.M. (2023). Exploiting deep learning in the performance evaluation of EH-based coordinated direct and relay transmission system with cognitive NOMA. *IEEE Communications Letters* 27 (6): 1501–1505. https://doi.org/10.1109/LCOMM.2023.3258665.

8 Singh, C.K., Upadhyay, P.K., Lehtomäki, J., and Juntti, M. (2024). Performance analysis with deep learning assay for cooperative UAV-borne IRS NOMA networks under non-ideal system imperfections. *IEEE Transactions on Vehicular Technology* 73 (1): 1065–1083. https://doi.org/10.1109/TVT.2023.3309619

9 Yuan, J., Liang, Y.C., Joung, J. et al. (2021). Intelligent reflecting surface-assisted cognitive radio system. *IEEE Transactions on Communications* 69 (1): 675–687. https://doi.org/10.1109/TCOMM.2020.3033006.

10 Chen, Y., Wang, Y., Zhang, J., and Di Renzo, M. (2021). QoS-driven spectrum sharing for reconfigurable intelligent surfaces (RISs) aided vehicular networks. *IEEE Transactions on Wireless Communications* 20 (9): 5969–5985. https://doi.org/10.1109/TWC.2021.3071332.

11 Kafafy, M., Ibrahim, A.S., and Ismail, M.H. (2022). Optimal placement of reconfigurable intelligent surfaces for spectrum coexistence with radars. *IEEE*

Transactions on Vehicular Technology 71 (6): 6574–6585. https://doi.org/10.1109/TVT.2022.3165391.

12 Yang, B., Cao, X., Huang, C. et al. (2022). Federated spectrum learning for reconfigurable intelligent surfaces-aided wireless edge networks. *IEEE Transactions on Wireless Communications* 21 (11): 9610–9626. https://doi.org/10.1109/TWC.2022.3178445.

13 Gautam, S., Lagunas, E., Chatzinotas, S., and Ottersten, B. (2019). Relay selection and resource allocation for SWIPT in multi-user OFDMA systems. *IEEE Transactions on Wireless Communications* 18 (5): 2493–2508.

14 Alevizos, P.N. and Bletsas, A. (2018). Sensitive and nonlinear far-field RF energy harvesting in wireless communications. *IEEE Transactions on Wireless Communications* 17 (6): 3670–3685. https://doi.org/10.1109/TWC.2018.2812889.

15 Boshkovska, E., Ng, D.W.K., Zlatanov, N., and Schober, R. (2015). Practical non-linear energy harvesting model and resource allocation for SWIPT systems. *IEEE Communications Letters* 19 (12): 2082–2085. https://doi.org/10.1109/LCOMM.2015.2478460.

16 Wu, Q., Guan, X., and Zhang, R. (2022). Intelligent reflecting surface-aided wireless energy and information transmission: an overview. *Proceedings of the IEEE* 110 (1): 150–170. https://doi.org/10.1109/JPROC.2021.3121790.

17 Sharma, P.K., Sharma, N., Dhok, S., and Singh, A. (2023). RIS-assisted FD short packet communication with non-linear EH. *IEEE Communications Letters* 27 (2): 522–526. https://doi.org/10.1109/LCOMM.2022.3223286.

18 Dhok, S., Raut, P., Sharma, P.K. et al. (2021). Non-linear energy harvesting in RIS-assisted URLLC networks for industry automation. *IEEE Transactions on Communications* 69 (11): 7761–7774. https://doi.org/10.1109/TCOMM.2021.3100611.

19 Do, T.N., Kaddoum, G., Nguyen, T.L. et al. (2021). Multi-RIS-aided wireless systems: statistical characterization and performance analysis. *IEEE Transactions on Communications* 69 (12): 8641–8658. https://doi.org/10.1109/TCOMM.2021.3117599.

20 Galappaththige, D.L., Kudathanthirige, D., and Amarasuriya, G. (2020). Performance analysis of distributed intelligent reflective surface aided communications. *GLOBECOM 2020 - 2020 IEEE Global Communications Conference*, 1–6. https://doi.org/10.1109/GLOBECOM42002.2020.9348102.

21 Yildirim, I., Uyrus, A., and Basar, E. (2021). Modeling and analysis of reconfigurable intelligent surfaces for indoor and outdoor applications in future wireless networks. *IEEE Transactions on Communications* 69 (2): 1290–1301. https://doi.org/10.1109/TCOMM.2020.3035391.

22 Aldababsa, M., Salhab, A.M., Nasir, A.A. et al. (2023). Multiple RISs-aided networks: performance analysis and optimization. *IEEE Transactions on Vehicular Technology* 72 (6): 7545–7559. https://doi.org/10.1109/TVT.2023.3242046.

23 Sharma, N. and Gautam, S. (2023). Optimizing RIS-assisted wireless communication systems with non-linear energy harvesting. *2023 5th International Conference on Energy, Power and Environment: Towards Flexible Green Energy Technologies (ICEPE)*, 1–5. https://doi.org/10.1109/ICEPE57949.2023.10201582.

24 Sharma, N. and Gautam, S. (2023). Throughput optimization for RIS-assisted SWIPT-IoTs with hardware impairments. *Proceedings of IEEE International Conference on Computing, Communication and Networking Technologies (ICCCNT).*

25 Sharma, N. (2023). On optimizing RIS-aided SWIPT-IoTs with power splitting-based non-linear energy harvesting. *GLOBECOM 2023 - 2023 IEEE Global Communications Conference*, 619–624. https://doi.org/10.1109/GLOBECOM54140.2023.10436799.

26 Sharma, N. and Gautam, S. (2024). Optimization strategy for RIS-assisted SWIPT-IoTs with non-linear energy harvesting. *2024 National Conference on Communications (NCC)*, 1–6. https://doi.org/10.1109/NCC60321.2024.10485849.

27 Wang, W., Ni, W., Tian, H., and Al-Dhahir, N. (2022). Multi-functional RIS: an integration of reflection, amplification, and energy harvesting. *GLOBECOM 2022 - 2022 IEEE Global Communications Conference*, 1–6. https://doi.org/10.1109/GLOBECOM48099.2022.10000917.

28 Xu, Y., Gu, B., Gao, Z. et al. (2024). Applying RIS in multi-user SWIPT-WPCN systems: a robust and environmentally-friendly design. *IEEE Transactions on Cognitive Communications and Networking* 10 (1): 209–222. https://doi.org/10.1109/TCCN.2023.3324636.

29 Mohamed, A., Zappone, A., and Di Renzo, M. (2022). Bi-objective optimization of information rate and harvested power in RIS-aided SWIPT systems. *IEEE Wireless Communications Letters* 11 (10): 2195–2199. https://doi.org/10.1109/LWC.2022.3196906.

30 Nwufo, C., Sun, Y., Simpson, O., and Cao, P. (2023). Secrecy energy efficiency maximization in multi-RIS-aided SWIPT wireless network. *2023 IEEE 97th Vehicular Technology Conference (VTC2023-Spring)*, 1–7. https://doi.org/10.1109/VTC2023-Spring57618.2023.10199952.

31 Galappaththige, D.L. and Baduge, G.A. (2022). Exploiting distributed IRSs for enabling SWIPT. *IEEE Wireless Communications Letters* 11 (4): 673–677. https://doi.org/10.1109/LWC.2021.3134630.

32 Le, N.P. and Le, K.N. (2022). Distributed-RIS assisted wireless powered communication networks in the finite blocklength regim. *IEEE Communications Letters* 26 (12): 2884–2888. https://doi.org/10.1109/LCOMM.2022.3205005.

33 Mohamed, Z. and Aïssa, S. (2022). Wireless power transfer aided with reconfigurable intelligent surfaces: design, and coverage analysis. *2022 IEEE 33rd Annual International Symposium on Personal, Indoor and Mobile Radio*

Communications (PIMRC), 695–700. https://doi.org/10.1109/PIMRC54779.2022.9977756.

34 Gautam, S., Solanki, S., Sharma, S.K. et al. (2021). Hybrid active-and-passive relaying model for 6G-IoT greencom networks with SWIPT. *Sensors* 21 (18): 6013.

35 Gautam, S., Lagunas, E., Bandi, A. et al. (2020). Multigroup multicast precoding for energy optimization in SWIPT systems with heterogeneous users. *IEEE Open Journal of the Communications Society* 1: 92–108. https://doi.org/10.1109/OJCOMS.2019.2962077.

36 Yang, M., Xu, Y., Huang, C. et al. (2022). Sum-rate maximization in RIS-aided wireless-powered D2D communication networks. *2022 IEEE 33rd Annual International Symposium on Personal, Indoor and Mobile Radio Communications (PIMRC)*, 307–312. https://doi.org/10.1109/PIMRC54779.2022.9977525.

37 Alves, H. and Lopez, O.A. (2021). Wireless RF energy transfer in the massive IoT era: towards sustainable zero-energy networks.

38 Xu, K., Shen, Z., Zhang, M. et al. (2019). Beam-domain SWIPT for mMIMO system with nonlinear energy harvesting legitimate terminals and a non-cooperative terminal. *IEEE Transactions on Green Communications and Networking* 3 (3): 703–720. https://doi.org/10.1109/TGCN.2019.2900343.

39 Alwazani, H. and Chaaban, A. (2020). Channel estimation for distributed intelligent reflecting surfaces assisted multi-user MISO systems. *2020 IEEE Globecom Workshops (GC Wkshps*, 1–6. https://doi.org/10.1109/GCWkshps50303.2020.9367461.

40 Series, P. (2017). Propagation data and prediction methods for the planning of short-range outdoor radiocommunication systems and radio local area networks in the frequency range 300 MHz to 100 GHz. *ITU Recommmendations*, 1411–1419.

41 Gautam, S., Kumar, S., Chatzinotas, S., and Ottersten, B. (2021). Experimental evaluation of RF waveform designs for wireless power transfer using software defined radio. *IEEE Access* 9: 132609–132622. https://doi.org/10.1109/ACCESS.2021.3115048.

42 Li, H., Shen, S., Nerini, M., and Clerckx, B. (2024). Reconfigurable intelligent surfaces 2.0: beyond diagonal phase shift matrices. *IEEE Communications Magazine* 62 (3): 102–108. https://doi.org/10.1109/MCOM.001.2300019.

43 Uniyal, S., Vishwakarma, N., Sharma, S., and Swaminathan, R. (2023). Intelligent reflecting surfaces-aided mixed FSO/RF communication system. *2023 IEEE Wireless Communications and Networking Conference (WCNC)*, 1–6. https://doi.org/10.1109/WCNC55385.2023.10118874.

44 Kokare, M.K.B., Swaminathan, R., and Gautam, S. (2024). Performance of multiple IRS-enabled V2V communication over double generalized gamma fading channel. *Proceedings of 2024 IEEE COMSNETS*, 913–919. https://doi.org/10.1109/COMSNETS59351.2024.10426981.

45 Kaushik, A., Singh, R., Dayarathna, S. et al. (2024). Toward integrated sensing and communications for 6G: key enabling technologies, standardization, and challenges. *IEEE Communications Standards Magazine* 8 (2): 52–59. https://doi.org/10.1109/MCOMSTD.0007.2300043.

7

Reinforcement Learning and Deep Learning-Assisted Spectrum Management for RIS-SWIPT-Enabled 6G Systems

Manojkumar B. Kokare, Purva Sharma, Swaminathan Ramabadran, Vimal Bhatia, and Sumit Gautam

Department of Electrical Engineering, Indian Institute of Technology Indore, Indore, Madhya Pradesh, India

7.1 Introduction

In recent years, autonomous driving and intelligent transportation system (ITS) applications have gained much attention from both industry and academia. ITS applications in the 6G mainly require low latency and high reliability while involving massive information transmission with ultrawide bandwidth. In the future, vehicle-to-everything (V2X) communication will play an important role in ITS, which enables vehicles to communicate with various entities within their environment, including other vehicles (V2V), roadside infrastructure (V2I), pedestrians (V2P), and even networks (V2N). This communication can be achieved through various wireless technologies, such as dedicated short-range communication (DSRC), and cellular vehicle-to-everything (C-V2X) [1]. Moreover, ITS has expanded from the ground to the air. To achieve high efficiency and reliability for a massive number of ground as well as aerial vehicles, it is mandatory to build high-speed, low-latency, and high-quality communication systems.

6G is projected to improve throughput by 1000 times over 5G by expanding the frequency spectrum to Terahertz (THz) range [2]. This ultrahigh frequency technology allows faster transmission and greater user access. However, high-frequency wave attenuation reduces transmission coverage and also increases the number of base stations (BS) and energy usage. In THz communication because of the short wavelength, physical objects can quickly impede the propagation link, introducing tight restrictions on beam alignment. Physical objects such as buildings in urban environments or hills and plants in the countryside may generate substantial shadowing effects in V2X communications

Intelligent Spectrum Management: Towards 6G, First Edition.
Edited by Sridhar Iyer, Anshuman Kalla, Onel Alcaraz López, and Chamitha De Alwis.

leading to a decrease in energy and spectrum efficiency. Furthermore, the high mobility of ground and aerial vehicles also affects channel stability.

The fundamental factors that impede high capacity and ultrareliable communications in 6G V2X systems are obstacles along with time-varying and random wireless channels. RISs/intelligent reflecting surfaces (IRSs) have emerged as a viable paradigm for reconfiguring the random radio/channel propagation environment to meet the 6G Quality-of-Service (QoS) requirements. A RIS comprises many passive reflecting elements that can dynamically alter the phase or amplitude of the incident signal to improve the wireless system performance. These surfaces can be densely placed in the wireless system to reconfigure their reflection to obtain the required propagation environment intelligently [3]. The authors in [4] offer an in-depth analysis of RISs for 6G communication and explores the various types of RISs, highlighting their advantages and limitations. Additionally, the chapter discussed the experimental setups used to evaluate RISs performance and presents the results. The discussion also includes potential future research areas and challenges within the 6G ecosystem, providing a comprehensive overview of the current state and future directions of RISs technology in next-generation communications. RISs establish a new propagation path sidestepping the obstacles, which increases the desired rate at the receiver and avoids the propagation loss. The RIS-assisted network provides dynamic control of the propagation environment, leading to a quantum leap in reliability and capacity.

6G communication systems are expected to reduce energy consumption and provide high-efficiency information transfer. Overall energy consumption drastically increases in wireless networks as the number of nodes increases. Harvesting energy from the wireless environment is a promising solution to reduce the consumption of energy from nonrenewable sources. Transmitted radio-frequency (RF) signals can be a new source for wireless power transfer (WPT) along with conventional wireless information transfer (WIT) [5]. SWIPT offers great convenience for transferring power and information simultaneously through a wireless medium. Thus, the receiver decodes the harvested energy and information from the received RF signal. Time switching (TS) or power splitting (PS) are the two protocols used for the SWIPT mechanism. In TS protocol, switching between time slots is done to harvest energy and information decoding. In PS protocol, received power is split in two and used to harvest energy and information decoding separately [6].

Authors in [7] optimize an IRS-assisted multiple-input multiple-output (MIMO)-orthogonal frequency division multiplexing (OFDM) SWIPT system with a nonlinear energy harvesting model to maximize data rates. They propose an alternating optimization (AO) algorithm for a high-quality suboptimal solution and a heuristic two-stage optimization algorithm for reduced complexity.

Simulations show the AO algorithm significantly improves data rates, while the heuristic algorithm performs comparably with lower complexity. In [8], an IRS-aided SWIPT system enhances transmission efficiency by compensating for RF signal loss. Using the block coordinate descent (BCD) method, the achievable rate-energy (R-E) region is optimized considering harvester nonlinearity. Simulations demonstrate significant R-E gains and robustness to channel state information at the transmitter (CSIT) inaccuracy and finite IRS states. A survey on IRS-aided wireless energy transmission (WET) and systems with both WIT and WET, such as SWIPT and wireless powered communication network (WPCN), is provided. The survey highlights design objectives, receiver architecture, and constraints, presenting state-of-the-art solutions and proposing new approaches and future research directions are discussed in [9].

Combining V2X with RISs for SWIPT can provide energy-efficient and high data rate services in dynamic situations, as RISs are reconfigurable, compatible, and easy to deploy. Harvesting energy at the receiver node ensures that RIS-assisted V2X communication lasts longer. SWIPT system collects energy from the received RF signal and utilizes it for further onboard energy requirements. The resource allocation for the RIS-assisted V2X SWIPT system can be formulated as the nonconvex optimization problem for maximization of rate and harvested energy using a nonlinear energy harvesting model.

This chapter aims to cover several relevant prospects in 6G wireless communication systems, including the applications of 6G in vehicular scenarios and the necessity of 6G in RIS-SWIPT-enabled wireless communication systems. Thus, the concept of RIS technology, specifications, and architecture are described to develop a basic understanding. The importance of SWIPT and the integration of RIS with SWIPT are also discussed. Moreover, in order to address the energy harvesting problem, an RIS-assisted V2V SWIPT system based on a TS scheme is proposed, where three optimization problems are solved: (i) optimization of power for maximization of rate and (ii) optimization of power for maximization of total harvested energy. Furthermore, some of the most relevant challenges and crucial research aspects related to RIS-SWIPT-enabled 6G systems are highlighted. The working of DRL framework to address the challenges of RIS-SWIPT-enabled 6G systems is also discussed. Moreover, we provide a discussion on the limitations of the existing technology that addresses the challenges and important future research directions using DRL methods.

The organization of this chapter is as follows. Section 5.2 discusses the concept, architecture, and characteristics of RIS technology. The concept of the proposed system model is described in Section 5.3. Section 5.4 elucidates the existing methods proposed to solve the key challenges, discusses the motivation behind using DRL methods in RIS-SWIPT-enabled 6G systems, and explains the workings of the DRL framework for RIS-SWIPT-enabled 6G systems. Section 5.5

presents open issues and challenges in RIS-SWIPT-enabled 6G systems. Finally, Section 5.6 concludes the chapter.

7.2 RIS Design and Characteristics

The RISs are two-dimensional (2D) reconfigurable sheets that modify environmental propagation to increase the signal quality at the receiver. The construction of RISs from a large number of inexpensive and passive components are capable of altering radio waves impinging on them in ways that natural materials cannot. Such, metamaterial-based RISs are called metasurfaces. These metasurfaces are placed between the transmitter and receiver, acting as reflection/refraction surfaces. Like other similar technologies, RISs do not require any complex encoding and decoding processing, and in addition, it does not require any power source. The hardware architecture of the RIS is as illustrated in Figure 7.1. Each metasurface element is embedded by a positive-intrinsic-negative (PIN)/Varactor diode. Generally, these diodes are turned on and off by an external bias yielding two possible states: reflection, and refraction.

The main characteristics of the RIS are listed as [10]:

- **Reconfigurable:** Every element of RIS is controlled to modify the amplitude, frequency, phase, and polarization of the incident wave.
- **Less Power Consumption:** Both active and passive RIS reconfigure the channel by the reflection of the signal and it uses minimal power to control the phases. Additionally, the RIS does not require analog-to-digital or digital-to-analog conversion and power amplification.

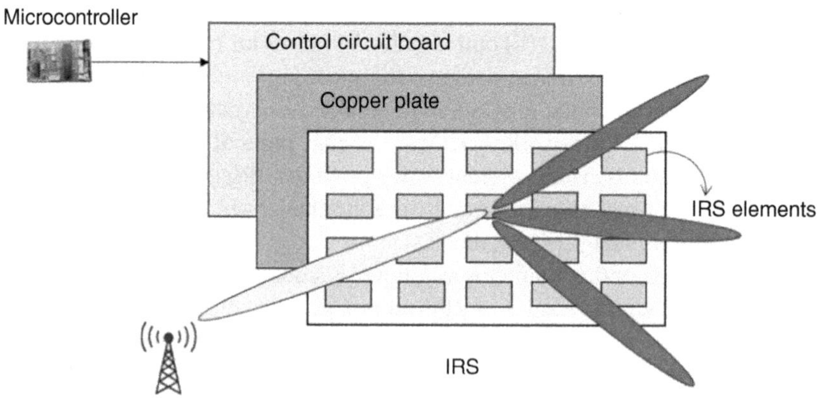

Figure 7.1 RIS architecture.

- **Easy to Deploy:** The deployment of RIS is easy due to the 2D architecture of surfaces and it can be deployed on walls of buildings, roadside poles, billboards, and aerial platforms due to very low cost.
- **Spectral Efficiency Enhancement:** In wireless communication systems power loss occurs if the transmitter and receiver are at a far distance. RIS can reconfigure the propagation environment by creating a virtual line of sight between the transmitter and receiver formed via a passive reflecting element, which gives enhancement in the signal-to-interference-plus-noise ratio (SINR).
- **Compatibility:** RIS-enabled networks are compatible with all existing networks hardware and standards.

7.3 SWIPT Protocols

In the TS scheme, the received signal is utilized at the destination vehicle to harvest energy for αT seconds and then perform information decoding for the remaining $(1-\alpha)T$ seconds. Here, we use the nonlinear Energy Harvesting (EH) model based on the sigmoidal function [11]. Hence the total harvested energy is expressed as:

$$E_h^{NL} = \frac{T'\epsilon}{1-\Phi}\left(\frac{1}{1+exp(-aP+ab)} - \Phi\right), \tag{7.1}$$

where $\Phi = \frac{1}{1+exp(ab)}$ is a constant for the zero-input/zero-output response of EH. a and b are constants related to circuit specifications, ϵ is the energy conversion efficiency of the receiver, and P is signal power. The harvested energy at the TS-SWIPT receiver is given by

$$E_h = \frac{\alpha T\epsilon}{1-\Phi}\left(\frac{1}{1+exp(-aP_t l_{ris}^2 Z^2+ab)} - \Phi\right). \tag{7.2}$$

Correspondingly, the overall rate at the receiver is given by

$$R_u = (1-\alpha)\log_2(1+\gamma_u). \tag{7.3}$$

As shown in Figure 7.2, we considered a RIS-aided V2I SWIPT system based on a TS scheme, where a single antenna roadside source (R_S) delivers information and power simultaneously to vehicle destination (V_D) through a RIS, which consists of N elements. We assume that the direct transmission between R_S and V_D is not possible and the transmission is done only via RIS. We consider $\overline{h}_r = |h_r|\, e^{j\theta_r}$ and $\overline{g}_r = |g_r|\, e^{j\xi_r}$ are the channel fading coefficients between R_S and r^{th} element of *RIS* and between the rth element of *RIS* and V_D, respectively. The fading channels are distributed according to a double generalized Gamma (dGG) distribution with shaping parameters $(\alpha_{1,r}, \beta_{1,r}, \alpha_{2,r}, \beta_{2,r})$ and $(\alpha_{3,r}, \beta_{3,r}, \alpha_{4,r}, \beta_{4,r})$, respectively. dGG efficiently captures the nonlinearities of the propagation medium, a common

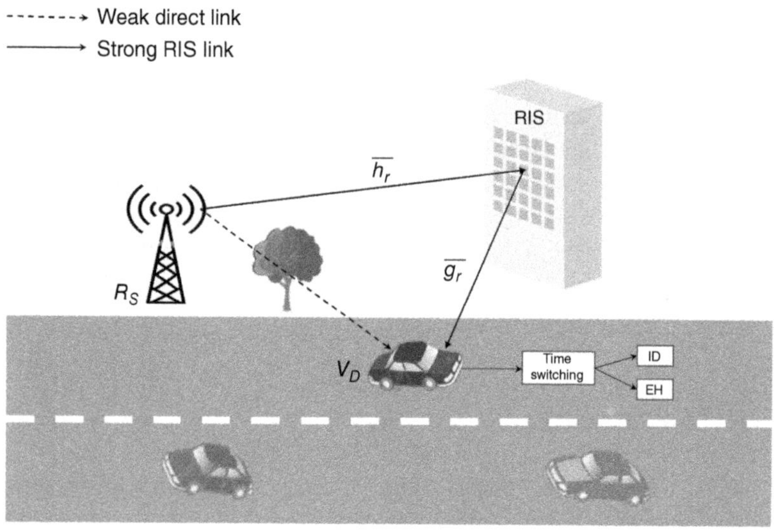

Figure 7.2 RIS-assisted SWIPT V2I communication.

occurrence in nonhomogeneous environments, and also includes important distributions such as double-Nakagami, Gamma-Gamma, double-Rayleigh, double-Weibull, and Weibull-Gamma.

The probability distribution function (PDF) of the dGG random variable (RV) $X(\alpha_1, \beta_1, \alpha_2, \beta_2)$ in terms of Fox's H-function is given by [12]

$$f_X(x) = \Theta x^{\alpha_2\beta_2-1} H_{0,2}^{2,0}\left(\phi x^{\alpha_2}\middle|_{(0,1),\left(\frac{\alpha_1\beta_1-\alpha_2\beta_2}{\alpha_1},\frac{\alpha_2}{\alpha_1}\right)}^{-}\right), \tag{7.4}$$

where $\alpha_1, \beta_1, \alpha_2$, and β_2 are the Gamma distribution shaping parameters, $\Theta = \frac{\alpha_2}{\left(\left(\frac{\Omega_1}{\beta_1}\right)^{\frac{\alpha_2\beta_2}{\alpha_1}}\left(\frac{\Omega_2}{\beta_2}\right)^{\beta_2}\Gamma(\beta_1)\Gamma(\beta_2)\right)}$, $\phi = \left(\frac{\beta_2}{\Omega_2}\right)\left(\frac{\beta_1}{\Omega_1}\right)^{\frac{\alpha_2}{\alpha_1}}$, and $\Omega_j = \left(\frac{\left(E\left[X_j^2\right]\right)\Gamma(\beta_1)}{\Gamma\left(\beta_j+\frac{2}{\alpha_j}\right)}\right)^{\frac{\alpha_j}{2}}$, $j = 1, 2$ is the α_jth root mean value.

The signal received at V_D by *RIS* is expressed as:

$$y = \sqrt{P_t l_{ris}}\left[\sum_{r=1}^{N}\overline{h}_r\overline{g}_r e^{j\phi_r}\right]x + n, \tag{7.5}$$

where P_t represents the average power of the transmitted signal, $e^{j\phi_r}$ is the phase shift of the rth reflecting element of the surface $\phi_r \in [0, 2\pi)$, x is the transmitted signal/symbol, n denotes additive white Gaussian noise (AWGN), which follows complex Gaussian distribution with zero-mean and variance N_0, that is, $n \sim \mathcal{CN}(0, N_0)$. l_{ris} is the free space path-loss, which is given by

$$l_{ris} = \frac{\lambda^2\sqrt{G_tG_r}}{16\pi d_1 d_2}, \tag{7.6}$$

where λ is the wavelength, G_t and G_r are the transmit and receive antenna gains respectively, and d_1 and d_2 represent the distance from R_S to RIS and RIS to V_D, respectively.

We consider the TS scheme for SWIPT with block transmission of duration T seconds. The received instantaneous SNR is expressed as:

$$\gamma = \frac{P_t l_{ris}^2 |\sum_{r=1}^{N} h_r g_r e^{j(\phi_r-\theta_r-\xi_r)}|^2}{N_0} = \frac{P_t l_{ris}^2 Z^2}{N_0}, \tag{7.7}$$

where $z^2 = \left|\sum_{r=1}^{N} h_r g_r\right|^2$. The optimal choice of ϕ_r that maximizes e2e SNR is $\phi_r = \theta_r + \xi_r$.

In what follows next, optimization problems are formulated to maximize the Rate and harvested energy for the RIS-aided SWIPT V2X system.

7.3.1 Rate Maximization via RIS

We consider the problem of optimization of power that maximizes the overall rate, while ensuring that the rate and harvested energy at the destination are, respectively, above a certain threshold, and the transmit power does not exceed a maximum limiting value P_{max}. Mathematically, the overall problem is expressed as:

$$\begin{aligned}
(P_1) : \underset{\alpha,P}{\text{maximize}} \quad & R_u \\
\text{subject to} \quad & (C_1) : R_u \geq R_{th}, \\
& (C_2) : E_h \geq \Xi_{th}, \\
& (C_3) : 0 \leq P \leq P_{max}, \\
& (C_4) : 0 \leq \alpha \leq 1,
\end{aligned} \tag{7.8}$$

where R_{th} and Ξ_{th} are the minimum rate and harvested energy demanded by the destination vehicle, respectively, and P_{max} is the upper limit of power. The problem in (P_1) is of a nonconvex form, due to joint parameter optimization of α and P. The constraint on R_u is considered in the problem to ensure the nonnegativity of the objective function. In this vein, obtaining suitable solutions within polynomial time is cumbersome. However, an iterative-method-based optimization wherein one parameter is kept constant while optimizing the other followed by the optimization of the former, keeping the latter (optimized) fixed and the corresponding process is repeated until the convergence of the objective [13]. Alternatively, a Lagrange dual function-based solution is also possible wherein the dual problem of (P_1) is formulated, which is always convex [14]. The iterative algorithm to solve the problem is given in Algorithm 7.1, which is detailed next.

Algorithm 7.1 Iterative maximization algorithm

Initialize α, R_{th}, Ξ_{th}, P_{max}
Repeat
Solve (P_1) or (P_2) for fixed α to obtain P^*
Solve (P_1) or (P_2) for fixed P^* to obtain α^*
find R_u or E_h using P^* and α^*, save $R(n)$ or $E_h(n)$ and $n = n + 1$
Until: $R(n) \geq R(n-1)$ or $E_h(n) \geq E_h(n-1)$

7.3.2 Maximization of Total Harvested Energy via RIS

Herein, we consider the problem of maximization of power that maximizes the total harvested energy, while ensuring that the rate and harvested energy at the destination are above the certain respective thresholds, and transmit power does not exceed the maximum value P_{max}. The inclusion of the constraint on E_h is necessary to guarantee that the objective function remains nonnegative. The above-mentioned problem is respectively expressed as:

$$
\begin{aligned}
(P_2):\ \underset{\alpha,P}{\text{maximize}}\quad & E_h \\
\text{subject to}\quad & (C_1):\ R_u \geq R_{th}, \\
& (C_2):\ E_h \geq \Xi_{th}, \\
& (C_3):\ 0 \leq P \leq P_{max}, \\
& (C_4):\ 0 \leq \alpha \leq 1,
\end{aligned} \tag{7.9}
$$

It is difficult to solve (P_2) directly as it turns into a nonconvex problem because of joint parameter optimization of α and P. By implying iterative method-based optimization given in Algorithm 7.1, which fixes one parameter and optimizes another and then fixes the latter to optimize the former sequentially until the convergence of the objective [15], we provide suitable solution(s) to address the aforementioned problem(s).

The above-mentioned algorithm may be utilized to find (suboptimal) solution for (P_1) or (P_2) within polynomial time. Other suitable methods may be employed to seek optimal solutions. In what follows next, we present the simulation results for the considered system setting and analyze the solutions obtained for (P_1) or (P_2) via Algorithm 7.1.

We evaluate the performance of the proposed RIS-assisted vehicular SWIPT system. We consider a signal of carrier frequency 6 GHz. The distances d_1 and d_2 are

2 and 4 m, respectively. The transmitter and the receiver antenna gain is 10 dBi. We set the maximum harvested power at each vehicle when the EH circuit is saturated as $\varepsilon = 10$ mW and we consider $a = 1500$ and $b = 0.0022$, which are obtained from curve fitting measurement data for nonlinear EH circuit [13]. The rate threshold value is considered as $R_{th} = 10$ bits/channel use and $P_{max} = 10$ W. The channel fading coefficients $\overline{h}_r$ and $\overline{g}_r$ are considered as identical with shaping parameters $(\alpha_{1,r} = \alpha_{2,r} = 1)$ and $(\beta_{1,r} = \beta_{2,r} = 2)$.

7.3.3 Maximized Rate versus Maximum Power

Figure 7.3 depicts the maximized rate versus maximum power at the receiver vehicle. We consider the thermal noise power as $N_O = -80$ dBW and the harvested energy threshold is considered as $\Xi_{th} = 1$ mJ. It is seen that the rate increases with the appreciating values of the transmitted power and it will be maximum at P_{max}. Moreover, it is seen that an increase in rate occurs when increases the number of RIS elements.

7.3.4 Maximized Harvested Energy Versus Maximum Power

Figure 7.4 shows the harvested energy versus maximum power at the receiver vehicle. With thermal noise power $N_O = -120$ dBW and a harvested energy threshold $\Xi_{th} = 1\mu$J, it is observed that harvested energy increases with transmitted

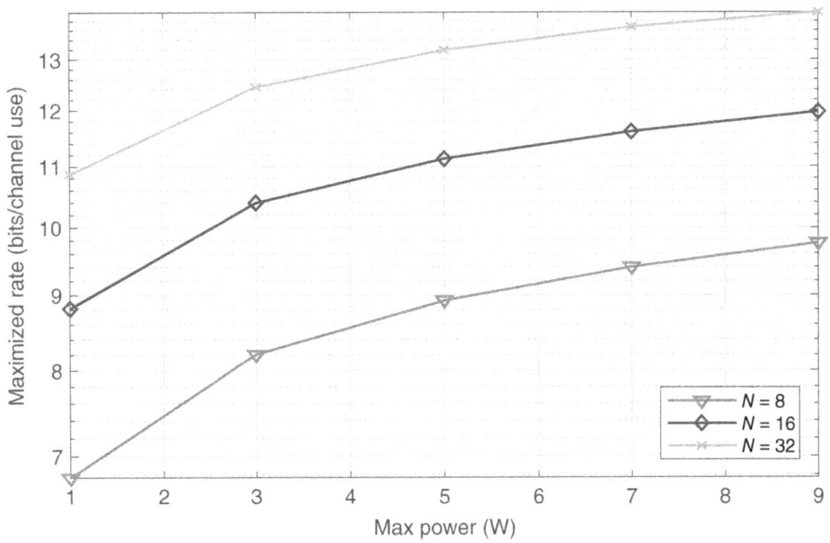

Figure 7.3 Rate (bits/channel use) versus maximum power (W).

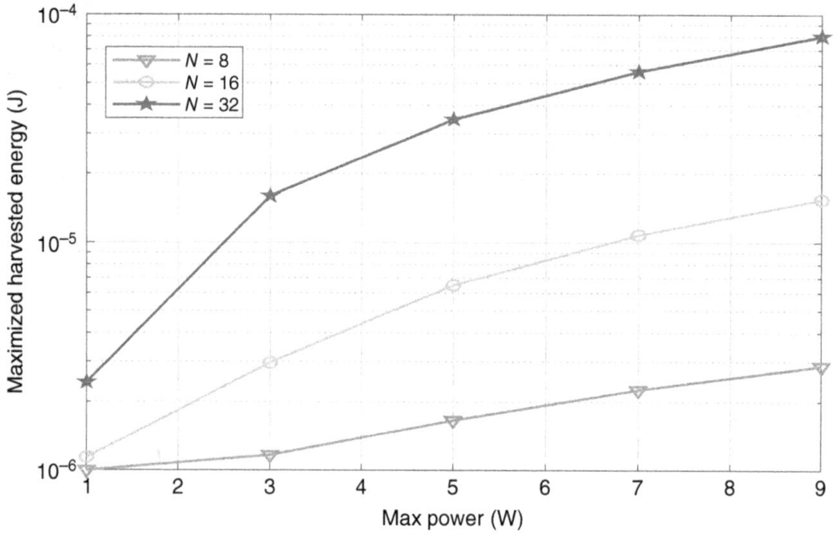

Figure 7.4 Harvested energy (Joules) versus maximum power (W).

power, reaching a maximum at P_{max}. Additionally, doubling the RIS elements from $N = 16$ to $N = 32$ reduces the power requirement by approximately 4.5 W to achieve 10 μJ of harvested energy.

7.4 DRL in RIS-Aided 6G Wireless Communication Systems

This section discusses a summary of the existing works, the motivation behind using DRL, and describes the working principle of the DRL framework to address the challenges of RIS-SWIPT-enabled 6G wireless communication systems.

7.4.1 State-of-the-Art and Motivation

Table 7.1 summarizes the existing works in the field of RIS-SWIPT-enabled 6G wireless communication systems. However, it has been observed from the literature that all the proposed methods are based on some predetermined rules to make decisions and may face difficulties in optimizing decisions in complex networking environments, such as RIS-SWIPT-enabled 6G wireless communication systems. Hence, inspired by the recent advances in artificial intelligence (AI)/machine learning (ML) methods that allow systems to learn from historical/previous data and make future predictions, the researchers have explored the capabilities of

Table 7.1 Summary of existing works that address various challenges in RIS-SWIPT enabled 6G wireless communication systems.

References	RIS	SWIPT	Description and key results
Han et al. [16]	✓		Ergodic spectrum efficiency is maximized by optimizing the phase shift of large intelligent surfaces (LIS)-assisted large-scale antenna system.
Deng and Pan [17]	✓	✓	External energy harvester eavesdropper harvests the energy and tries to eavesdrop the legitimate user information. The optimization problem to maximize the secrecy rate under the minimum required harvested energy is proposed.
Chen et al. [18]	✓		RIS surfaces are used to increase the SNR at the legit user in the multiple-input-single-output (MISO) system. Minimum-secrecy-rate maximization problem proposed with constraints on the reflecting coefficients.
Wu and Zhang [19]	✓	✓	Weighted sum rate maximization is studied for the SWIPT system assisted by RIS considering perfect CSI. It is verified that the rate-energy tradeoff is drastically improved with the assistance of RIS
Jaafar et al. [20]	✓	✓	Energy is harvested for both RIS and IoT nodes. To maximize the harvested energy and rate, time switching and phase shifting are controlled and the nonconvex problem is solved using a sequential approach.
Chen et al. [21]	✓		Resource allocation for vehicular network based on slow varying large-scale fading model is studied. Maximization of V2I link capacity considering minimum V2V SINR. The RIS reflection coefficients, power allocation, and spectrum allocation are jointly optimized using an alternating optimization algorithm.
Thien et al. [22]	✓	✓	A PS protocol is considered for IRS-assisted secure transmission maximization for SWIPT systems. The optimization problem is formulated to maximize the systems secrecy rate by finding the RIS phase shifts, PS factor, and optimal transmitter power. Also, to improve computation performance, a DL-based approach was proposed.

AI/ML in the field of wireless communications to improve the efficiency and performance of the system. The methods based on ML require labeling of data to identify the raw data and add informative labels that help the ML model to learn. However, as the dynamicity of the network increases (the information regarding network layout and channel states is unavailable), such ML-based methods do not work properly. Hence, DRL technology [23], a branch of AI/ML, has attracted researchers' attention in the networking field. DRL is a combination of deep learning (DL) and reinforcement learning (RL) technologies. DL builds on deep neural networks to create a complex connection between input and output data. Whereas, RL enables agents to learn complex behaviors and make perceptions in challenging networking environments [24]. The complete working operation of the DRL framework to address the challenges of RIS-SWIPT-enabled 6G wireless communication systems is discussed in Section 7.4.2.

7.4.2 DRL Framework for RIS-Assisted 6G Wireless Systems

This subsection describes the working principle of the DRL framework that can address various challenges such as resource assignment, optimization for phase shift, power allocation, passive/active beamforming, and many more in RIS-SWIPT-enabled 6G wireless communication systems.

Figure 7.5 illustrates the framework of DRL to solve the complex decision-making problem in an RIS-SWIPT-enabled 6G wireless communication environment. The DRL algorithm consists of modeling and training phases. The modeling phase has three stages: state *(S)*, action *(A)*, and reward *(R)* [25].

(1) **State:** The state *(S)* contains the complete environment's information, which includes the channel information of all users, the secrecy rate, the transmission data rate, etc. The customized state at time instant t is denoted by $s_t, s_t \in S$. The environment then sends the customized state (s_t) to the agent.
(2) **Action:** Action space *(A)* is a group of actions that are the response of an agent during the DRL process. Based on the observed state (s_t), every time an agent takes an action $a_t \in A$.
(3) **Reward:** Reward *(R)* is feedback to the agent from an environment. At every time t, the agent will receive a reward r_t, where $r_t \in R$, for performing the right action, or else be penalized.

In this DRL framework, the RIS-SWIPT-enabled 6G wireless communication system is treated as an environment. When the process starts, the central controller fetches the environment's state information and sends it to the agent to generate a customized state (s_t). The neural network of the DRL reads s_t and takes an action (a_t). The agent selects an action according to the output policy $\pi_t(A|s_t, \theta)$ (based on the specific optimization problem). After receiving feedback from the previous operation or action, a reward system produces an immediate reward (r_t). The agent collects and stores each experience as a tuple $E\{s_t, a_t, r_t, \gamma\}$, where $\gamma \in [0, 1]$ is a discount factor, in an experience buffer. This experience buffer $E\{s_t, a_t, r_t, \gamma\}$ is then used to train the learning model of the agent. The objective of the DRL framework is to learn the optimal policy based on the specific complex decision-making problem of the RIS-SWIPT-enabled 6G wireless communication system and maximize the cumulative reward in the long term. The different DRL

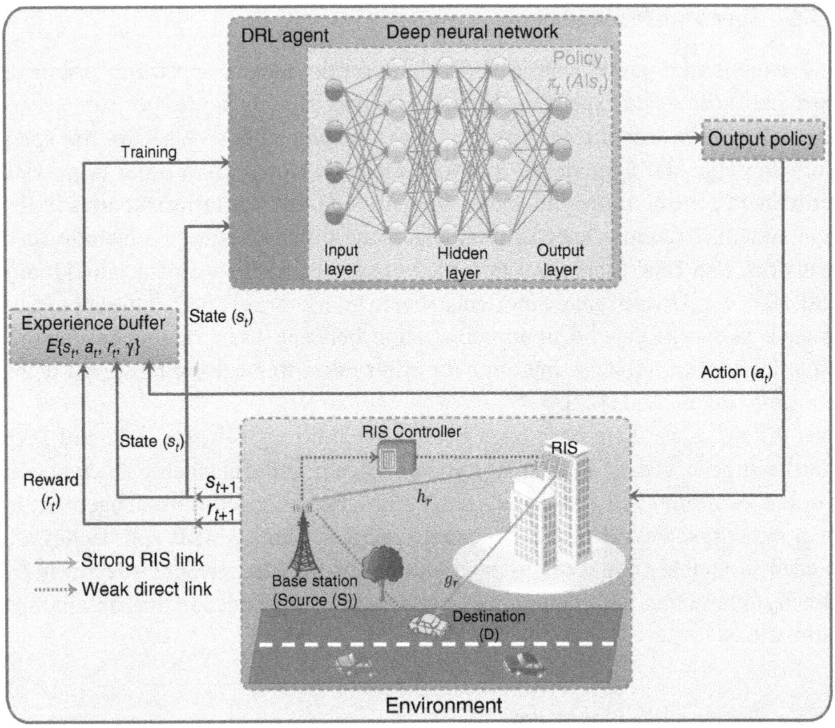

Figure 7.5 An illustration of DRL framework for solving research challenges of RIS-aided 6G wireless systems.

algorithms, such as deep Q-network (DQN), deep deterministic policy gradient (DDPG), twin delayed DDPG (TD3), trust region policy optimization (TRPO), and proximal policy optimization (PPO), can be applied to RIS-SWIPT-enabled 6G wireless communication systems to improve the performance.

7.5 Open Issues and Challenges

6G communication systems are fast evolving and are expected to overcome the limitations of the existing 5G communication systems in terms of data rate, reliability, processing, latency, and availability. Significant research efforts are needed to improve both the technical aspects of 6G as well as use cases of the 6G ecosystem. This section highlights some of the important open issues and challenges that are necessary to be addressed in the near future for RIS-SWIPT-enabled 6G wireless communication systems.

7.5.1 Spectrum Management

The exponential growth in mobile data traffic makes spectrum resources extremely scarce and expensive. Hence, the concept of dynamic spectrum access (DSA) has been introduced. The DSA has the capability to alleviate the spectrum shortage and upgrade the network capacity. However, in order to provide centralized control to the end users, there are no powerful infrastructures in the DSA system. Therefore, spectrum management is a challenging problem in such networks, and DSA users have to carry out spectrum management individually [26]. Hence, DSA users may experience severe interference. The interference issue in such networks depend upon cooperation between users or precise channel state information (CSI) estimations for other users that cannot be solved using conventional methods [27].

In recent times, there have been notable breakthroughs in ML, DL, and DRL that have been widely studied to improve system performance or efficiency in wireless communication systems. Hence, for efficient spectrum management in such networks, several new DRL-based spectrum management approaches are needed to enable DSA users to carry out spectrum management effectively by directly interacting with complex environments without depending on channel estimations.

7.5.2 Optimal RIS Placement

The optimized placement of RIS has received considerable research interest as it can significantly improve the performance of RIS-SWIPT-enabled 6G wireless communication systems. In RIS-SWIPT-enabled 6G wireless communication systems, the optimal placement of RIS is important to extend the RIS coverage area, enhance the signal-to-noise ratio level, and minimize the interference between the adjacent devices. With a high optimal achievable rate, the RIS placement can determine the efficiency and the optimal level of the reflected and distributed signal, respectively [28].

Several heuristic methods have been developed in the literature to optimize RIS placement in single and multi-RIS-aided communication systems. However, heuristic methods may face difficulties, as they rely on predetermined rules, in optimizing the placement decisions of the RIS in a complex environment. Therefore, in order to address the above-mentioned problem, ML/DL/RL has the capability to learn directly from experiences, make decisions, and optimize the placement of RIS for better performance in RIS-SWIPT-enabled 6G wireless communication system. Hence, efforts are needed to develop several optimal RIS placement strategies using DRL algorithms.

7.5.3 Channel Estimation

RIS channel estimation is crucial for controlling wireless propagation environments in RIS-aided wireless communication systems. The RIS-aided network must be aware of the CSI of the channels entering and leaving the RIS to perform beamforming correctly and choose the appropriate reflection pattern [29]. However, accurate channel estimation is challenging due to the lack of active components for baseband signal processing and the complexity of RIS-associated channels. Different channel estimation techniques are present in the literature, such as separate and cascaded channel estimation. These approaches have their pros and cons, with their effectiveness depending on signal processing, channel models, training designs, and hardware constraints [30]. Combining these techniques reduces real-time training overhead and enhances applicability for both time division duplexing (TDD) and frequency division duplexing (FDD) systems.

Recent research has explored various signal processing methods for RIS channel estimation have been explored, including classical channel estimation, full-ON RIS training reflection pattern, and training protocols. Classical channel estimation is widely applied in RIS-aided systems due to its low complexity. Compressed sensing is a promising tool for RIS channel estimation in high-frequency communication systems, exploiting the sparsity of RIS-associated channels and reducing training overhead [31].

It has been observed from the above studies that the channel model in RIS-aided communication becomes more complex due to the presence of multiple reflections and interactions between the incident signal, the reflecting elements on the surface, and the receiver. Traditional methods of channel estimation may struggle to convey these complexities with precision. ML algorithms, especially DL models, excel at capturing intricate data relationships, making them ideal for modeling these complex behaviors. Additionally, DRL algorithms enhance channel estimation accuracy in dynamic network by enabling the agent to learn and adapt to changes. Hence, DRL algorithms have also been applied to improve the channel estimation accuracy of the dynamic networking environment, which enables the DRL agent to learn and adapt to changing networks.

7.5.4 RIS Selection

Particularly, the integration of multiple reflecting paths produced by multiple RISs can greatly improve the wireless systems performance [32]. Hence, multiple RISs have been used in addition to a single RIS to assist with wireless communication systems. However, using all the RIS in the system can lead to high resource usage and increase overall system complexity. Therefore, in order to reduce resource usage, the RIS selection technique has been proposed. The RIS selection technique chooses the RIS with the highest SNR to aid communication to achieve low-complexity transmission [33], and the most suitable RIS can participate in wireless communication. Hence, efforts are required to develop several DRL-based RIS selection strategies to select the best option among multiple RISs to save the system resources and minimize its complexity.

7.5.5 Security and Privacy

Physical layer security (PLS) has emerged as an effective and alternative solution to conventional cryptography techniques for wireless communication systems that securely exchange information in the presence of an eavesdropper. Various approaches have been proposed in the literature to enhance the security of wireless communication systems [34]. However, in PLS systems, the deployment of numerous active antennas and relays increases the system complexity and hardware cost. Hence, in order to overcome the drawbacks of the existing approaches, achieve high energy efficiency, and improve the secrecy performance of wireless communication systems, RIS is proposed as a promising technique.

Initially, a simple model with single-antenna users and eavesdroppers was used to maximize the secrecy rate in RIS-aided communication systems via alternative optimization. This was extended to multiple antenna eavesdroppers and MIMO channels for secure communication between multiantenna BS and users [35]. Furthermore, a RIS-aided MIMO channel is considered, where information is securely exchanged between multiantenna BS to multiantenna legitimate user in the presence of an eavesdropper [36]. Traditional optimization techniques are inefficient for complex, large-scale systems. Thus, recent advances in AI/ML, such as DL and RL, can be exploited in dynamic RIS-aided secure communication systems to optimize the transmit and reflect beamforming in BS and RIS, respectively, under multiple eavesdroppers. Such DRL-based algorithms improve complex environments learning efficiency and secrecy performance by fully exploiting the environmental information. This makes DRL practically effective in dynamic and complex networks.

7.6 Conclusion

This chapter provided an overview of RIS-SWIPT-enabled 6G systems. Several interesting concepts required to build an understanding of RIS-SWIPT-enabled 6G systems have been covered, including the RIS architecture and its characteristics,

SWIPT, and a summary of existing methods to solve the important problems in RIS-SWIPT-enabled 6G systems. Moreover, a RIS-aided V2V SWIPT system based on TS scheme has been proposed to address the energy harvesting problem. Furthermore, several unexplored challenges have also been highlighted, which require further investigation. Moreover, a motivation behind using DRL in RIS-SWIPT-enabled 6G systems and a DRL framework have been discussed in order to solve the critical challenges of such as spectrum management. In the future, efficient DRL-based methods need to be developed for the already explored and new challenges of RIS-SWIPT-enabled 6G systems.

References

1 Osorio, D.P.M., Ahmad, I., Sánchez, J.D.V. et al. (2022). Towards 6G-enabled internet of vehicles: security and privacy. *IEEE Open Journal of the Communications Society* 3: 82–105. https://doi.org/10.1109/OJCOMS.2022.3143098.

2 Yastrebova, A., Kirichek, R., Koucheryavy, Y. et al. (2018). Future networks 2030: architecture & requirements. *2018 10th International Congress on Ultra Modern Telecommunications and Control Systems and Workshops (ICUMT)*, 1–8. https://doi.org/10.1109/ICUMT.2018.8631208.

3 Basar, E., Di Renzo, M., De Rosny, J. et al. (2019). Wireless communications through reconfigurable intelligent surfaces. *IEEE Access* 7: 116753–116773. https://doi.org/10.1109/ACCESS.2019.2935192.

4 Di Renzo, M. and Aravanis, A.I. (2022). Catching the 6G wave by using metamaterials: a reconfigurable intelligent surface paradigm. In: *Shaping Future 6G Networks: Needs, Impacts, and Technologies*, 69–87. https://doi.org/10.1002/9781119765554.ch6.

5 López, O.L.A., Alves, H., Souza, R.D. et al. (2021). Massive wireless energy transfer: enabling sustainable IoT toward 6G era. *IEEE Internet of Things Journal* 8 (11): 8816–8835. https://doi.org/10.1109/JIOT.2021.3050612.

6 Gautam, S. and Ubaidulla, P. (2017). Relay selection and transceiver design for joint wireless information and energy transfer in cooperative networks. *2017 IEEE 85th Vehicular Technology Conference (VTC Spring)*, 1–5. https://doi.org/10.1109/VTCSpring.2017.8108246.

7 Peng, X., Wu, P., Tan, H., and Xia, M. (2022). Optimization for IRS-assisted MIMO-OFDM SWIPT system with nonlinear EH model. *IEEE Internet of Things Journal* 9 (24): 25253–25268.

8 Zhao, Y., Clerckx, B., and Feng, Z. (2022). IRS-aided SWIPT: joint waveform, active and passive beamforming design under nonlinear harvester model. *IEEE Transactions on Communications* 70 (2): 1345–1359.

9 Wu, Q., Guan, X., and Zhang, R. (2022). Intelligent reflecting surface-aided wireless energy and information transmission: an overview. *Proceedings of the IEEE* 110 (1): 150–170. https://doi.org/10.1109/JPROC.2021.3121790.

10 López, O.L.A. Rosabal, O.M., Ruíz-Guirola, D.E. et al. (2023). Energy-sustainable IoT connectivity: vision, technological enablers, challenges, and future directions. *ArXiv*, abs/2306.02444. https://api.semanticscholar.org/CorpusID:259076049.

11 Boshkovska, E., Ng, D.W.K., Zlatanov, N., and Schober, R. (2015). Practical non-linear energy harvesting model and resource allocation for SWIPT systems. *IEEE Communications Letters* 19 (12): 2082–2085. https://doi.org/10.1109/LCOMM.2015.2478460.

12 Kokare, M.K.B., Swaminathan, R., and Gautam, S. (2024). Performance of multiple IRS-enabled V2V communication over double generalized gamma fading channel. *Proceedings of 2024 IEEE COMSNETS*, 913–919. https://doi.org/10.1109/COMSNETS59351.2024.10426981.

13 Sharma, N. and Gautam, S. (2023). Optimizing RIS-assisted wireless communication systems with non-linear energy harvesting. *2023 5th International Conference on Energy, Power and Environment: Towards Flexible Green Energy Technologies (ICEPE)*, 1–5.

14 Boyd, S. and Vandenberghe, L. (2004). *Convex Optimization*. Cambridge university press.

15 Sharma, N. and Gautam, S. (2023). Throughput optimization for RIS-assisted SWIPT-IoTs with hardware impairments. *2023 14th International Conference on Computing Communication and Networking Technologies (ICCCNT)*, 1–6. IEEE.

16 Han, Y., Tang, W., Jin, S. et al. (2019). Large intelligent surface-assisted wireless communication exploiting statistical CSI. *IEEE Transactions on Vehicular Technology* 68 (8): 8238–8242. https://doi.org/10.1109/TVT.2019.2923997.

17 Deng, Z. and Pan, Y. (2021). Optimal beamforming for IRS-assisted SWIPT system with an energy-harvesting eavesdropper. *Electronics* 10 (20): 2536. https://doi.org/10.3390/electronics10202536.

18 Chen, J., Liang, Y.-C., Pei, Y., and Guo, H. (2019). Intelligent reflecting surface: a programmable wireless environment for physical layer security. *IEEE Access* 7: 82599–82612. https://doi.org/10.1109/ACCESS.2019.2924034.

19 Wu, Q. and Zhang, R. (2020). Weighted sum power maximization for intelligent reflecting surface aided SWIPT. *IEEE Wireless Communications Letters* 9 (5): 586–590. https://doi.org/10.1109/LWC.2019.2961656.

20 Jaafar, W., Bariah, L., Muhaidat, S., and Yanikomeroglu, H. (2022). Time-switching and phase-shifting control for RIS-assisted SWIPT communications. *IEEE Wireless Communications Letters* 11 (8): 1728–1732. https://doi.org/10.1109/LWC.2022.3178630.

21 Chen, Y., Wang, Y., Zhang, J., and Li, Z. (2020). Resource allocation for intelligent reflecting surface aided vehicular communications. *IEEE Transactions on Vehicular Technology* 69 (10): 12321–12326. https://doi.org/10.1109/TVT.2020.3010252.

22 Thien, H.T., Tuan, P.-V., and Koo, I. (2022). A secure-transmission maximization scheme for SWIPT systems assisted by an intelligent reflecting surface and deep learning. *IEEE Access* 10: 31851–31867.

23 Sutton, R.S. and Barto, A.G. (2018). *Reinforcement Learning: An Introduction.* MIT Press.

24 Wang, W. and Zhang, W. (2022). Intelligent reflecting surface configurations for smart radio using deep reinforcement learning. *IEEE Journal on Selected Areas in Communications* 40 (8): 2335–2346.

25 Puspitasari, A.A. and Lee, B.M. (2023). A survey on reinforcement learning for reconfigurable intelligent surfaces in wireless communications. *Sensors* 23 (5): 2554.

26 Jabbari, B., Pickholtz, R., and Norton, M. (2010). Dynamic spectrum access and management [dynamic spectrum management]. *IEEE Wireless Communications* 17 (4): 6–15.

27 Song, H., Fang, X., and Fang, Y. (2016). Unlicensed spectra fusion and interference coordination for LTE systems. *IEEE Transactions on Mobile Computing* 15 (12): 3171–3184.

28 Perović, N.S., Tran, L.-N., Di Renzo, M., and Flanagan, M.F. (2021). Achievable rate optimization for MIMO systems with reconfigurable intelligent surfaces. *IEEE Transactions on Wireless Communications* 20 (6): 3865–3882.

29 Mutlu, U. and Kabalci, Y. (2023). Deep learning aided channel estimation in intelligent reflecting surfaces. *2023 5th Global Power, Energy and Communication Conference (GPECOM)*, 513–518. IEEE.

30 Zheng, B., You, C., Mei, W., and Zhang, R. (2022). A survey on channel estimation and practical passive beamforming design for intelligent reflecting surface aided wireless communications. *IEEE Communications Surveys & Tutorials* 24 (2): 1035–1071.

31 Lin, T., Yu, X., Zhu, Y., and Schober, R. (2022). Channel estimation for IRS-assisted millimeter-wave MIMO systems: sparsity-inspired approaches. *IEEE Transactions on Communications* 70 (6): 4078–4092.

32 Do, T.N., Kaddoum, G., Nguyen, T.L. et al. (2021). Multi-RIS-aided wireless systems: statistical characterization and performance analysis. *IEEE Transactions on Communications* 69 (12): 8641–8658. https://doi.org/10.1109/TCOMM.2021.3117599.

33 Aldababsa, M., Salhab, A.M., Nasir, A.A. et al. (2023). Multiple RISs-aided networks: performance analysis and optimization. *IEEE Transactions on Vehicular Technology* 72 (6): 7545–7559. https://doi.org/10.1109/TVT.2023.3242046.

34 Li, Q. and Yang, L. (2019). Beamforming for cooperative secure transmission in cognitive two-way relay networks. *IEEE Transactions on Information Forensics and Security* 15: 130–143.

35 Guan, X., Wu, Q., and Zhang, R. (2020). Intelligent reflecting surface assisted secrecy communication: is artificial noise helpful or not? *IEEE Wireless Communications Letters* 9 (6): 778–782.

36 Dong, L. and Wang, H.-M. (2020). Secure MIMO transmission via intelligent reflecting surface. *IEEE Wireless Communications Letters* 9 (6): 787–790.

8

RIS-Aided Low Complexity Waveform Design for Joint Sensing and Communications

Christos Tsinos[1], *Soumya P. Dash*[2], *Aryan Kaushik*[3], *Aakash Arora*[4], *and Marco Di Renzo*[5]

[1]*Department of Digital Industry Technologies, National and Kapodistrian University of Athens, Athens, Greece*
[2]*School of Electrical Sciences, Indian Institute of Technology Bhubaneswar, Bhubaneswar, India*
[3]*School of Engineering and Informatics, University of Sussex, Brighton, UK*
[4]*Radio Systems Group, Faculty of Electrical Engineering, Mathematics and Computer Science, University of Twente, Enschede, The Netherlands*
[5]*Université Paris-Saclay, CNRS, CentraleSupélec, Laboratoire des Signaux et Systèmes, Gif-sur-Yvette, France*

8.1 Introduction

Projected subscriptions of the fifth generation (5G) are anticipated to hit 5 billion by 2028, indicating a faster rate of adoption compared to previous generation standards. This rapid growth can be attributed to the extensive deployment of the initial 5G standard [1]. Various countries are leveraging three distinct 5G spectrum bands, namely low-band (below 7 GHz, using frequency division duplex), mid-band (below 7 GHz, employing time division duplex), and high-band (above 24 GHz, utilizing millimeter wave (mmWave)), based on the availability of spectrum resources in their respective regions. For instance, the United States is making strides in all three bands; the United Kingdom focuses on the low and mid-bands, while South Korea concentrates solely on the mid-band [1]. The availability of spectrum resources holds paramount significance in the deployment of upcoming wireless standards such as 5G Advanced and the sixth generation (6G). The most recent Release 18 from the Third-Generation Partnership Project (3GPP) underscores enhancements in new radio (NR) mobility, radio frequency (RF), and dynamic spectrum sharing (DSS) within the NR framework [2]. The design of 5G NR allows for a unified and scalable air interface that

Intelligent Spectrum Management: Towards 6G, First Edition.
Edited by Sridhar Iyer, Anshuman Kalla, Onel Alcaraz López, and Chamitha De Alwis.

facilitates the coexistence of diverse 5G device classes. This design has been expanded to support the 60 GHz unlicensed band, with a priority on extending mmWave to 71 GHz in Release 17 [3].

Technologies operating at high frequencies, including mmWave (ranging from 30 to 300 GHz), sub-terahertz (THz), and THz, provide significantly broader bandwidth compared to earlier generation standards, aiming to keep pace with the continuously increasing demands of users. However, high-frequency channel models pose several challenges, such as substantial path loss, blockage effects, and predominantly Line-of-Sight (LoS) communications [4–6]. Conversely, the advantages highlight the ability to meet spectrum requirements for service providers, enabling them to support high channel capacity and rapid data rates. In the initial stages of 5G rollouts, the global community expressed concerns about spectrum availability, given the heavy congestion in the currently utilized RF/sub-6 GHz spectrum. As the projection for mobile data traffic shows a significant increase, sophisticated solutions for spectrum sharing become imperative to effectively utilize the available spectral resources [7]. Simultaneously, it is crucial to thoroughly explore the primary use cases of ultrareliable low-latency communication (URLLC), enhanced mobile broadband (eMBB), and massive machine-type communication (mMTC), while also meeting the demands for spectrum sharing in the upcoming generation of wireless standards.

While 5G Advanced and 6G demand extensive connectivity and high-speed data transmission, the current emphasis lies on the imperative requirement for cost-effective and energy-efficient solutions [8]. In the context of 6G mobile communications, attaining elevated data rates and engaging in long-term spectrum planning are envisioned as pivotal technological pursuits. Additionally, intelligently reusing existing hardware in conjunction with the concept of spectrum sharing stands as a crucial solution for the forthcoming generation of networks. The United Nations (UN) sustainable development goals, encompassing objectives such as "affordable and clean energy," "sustainable cities and communities," and "industry, innovation, and infrastructure," are set to influence the cost-effectiveness and sustainability of next-generation wireless technologies through the shared utilization of resources, including spectrum and hardware. Thus, as the demand for higher data rates, improved coverage, and efficient spectrum utilization continues to grow, traditional communication infrastructure faces limitations [9]. Over recent years, the use of reconfigurable intelligent surfaces (RIS) [10–17] and joint sensing and communications (JSAC) [18–25] have become groundbreaking concepts to emerge as a promising solution to overcome these challenges and reshape the landscape of wireless communications. As ubiquitous connectivity is one of the key usage scenarios in IMT-2030/6G framework, RIS based communication can be also utilized in nonterrestrial networks [26, 27] and further extensions in RIS model, such as with index modulation, can lead to advanced technical solutions for 6G [28]. Moreover, waveform design is crucial for such integrated sensing and communication (ISAC) systems due to its direct impact on system performance and efficiency. An effective waveform must seamlessly support both sensing and communication functionalities without compromising either. This involves optimizing parameters like bandwidth, modulation schemes, and signal structure to achieve high-resolution sensing while maintaining robust data transmission. The dual-use waveforms must be adaptable to different environments and applications, ensuring reliable performance under varying conditions. Properly designed waveforms can enhance spectrum utilization, reduce interference, and improve the accuracy of target detection and data delivery.

Therefore, drawing inspiration from these technological advancements, this chapter introduces an innovative analog architecture tailored for multiple-input multiple-output (MIMO) joint radar-communication (JRC) systems with and without RIS. This architecture hinges on a single-phase shifter per antenna and a shared variable gain amplifier (VGA) responsible for driving the phase-shifting network. Subsequently, we devise an optimal transmit waveform by formulating an optimization problem. This problem seeks to minimize multiuser interference (MUI) in the communication aspect while concurrently imposing specific radar beam patterns and adhering to constraints associated with the proposed architecture, notably per-antenna constant-modulus constraints. It is essential to note that this optimization problem is nonconvex, posing significant computational challenges, and remains an unaddressed issue in existing literature. To tackle this, we employ an efficient algorithm rooted in the primal-dual framework, as detailed in [29]. The numerical results we present demonstrate substantial gains in spectral efficiency within the MIMO JRC system compared to conventional fully digital architectures, all while maintaining favorable radar beam pattern performance. Furthermore, we evaluate the effect of using RIS on the performance of the system.

The rest of the chapter is organized as follows. Section 8.2 provides a description of RIS and ISAC systems and sheds light on their integration for the next-generation wireless systems. Section 8.3 provides a short survey of the necessity of waveform design for ISAC systems and outlines the different types of waveform designs provided in the literature. Section 8.4 considers a case study of the design of waveform design of a RIS-aided MIMO JRC system wherein the details the system model for the MIMO JRC communication system are detailed. The optimal waveform design for the non-RIS-assisted system and the corresponding algorithm are also described in this section, which is followed by the corresponding waveform design for the RIS-assisted system and its performance study. The concluding remarks of the chapter are provided in Section 8.5.

Notation: $\mathbf{A}$, $\mathbf{a}$, a: matrix, vector and scalar variable, respectively; $\text{tr}(\cdot)$, $\text{re}(\cdot)$, $(\cdot)^T$, $(\cdot)^H$, $(\cdot)^{-1}$, $(\cdot)^{-H}$, $||(\cdot)||_p$ and $||(\cdot)||_F$ denote the trace, real-term, transpose, Hermitian transpose, inverse, inverse Hermitian transpose, p norm and Frobenius norm, respectively; $\mathbf{A}_{i,j}$ and $\mathbf{a}_i$ are the (i,j)th and ith elements in $\mathbf{A}$ and $\mathbf{a}$, respectively; $|\cdot|$ is the modulus of a complex number. $\mathbf{I}_N$ is N-size identity matrix; $\mathbb{C}$ and $\mathbb{R}_+$ and denote the set of complex and real positive numbers, respectively; $\text{diag}\{\cdot\}$ is the operator that returns the diagonal entries of a matrix in diagonal matrix form; $\mathbb{E}(.)$ denotes the expectation operator; $\mathcal{CN}(a, b)$ is a complex Gaussian vector with mean a and variance b.

8.2 RIS and ISAC

8.2.1 Reconfigurable Intelligent Surfaces (RIS)

RIS, also known as smart surfaces, encompasses a new paradigm in wireless communication technology [30–32]. Unlike traditional antennas and base stations (BSs), which transmit and receive signals directly, RIS operates as passive scattering elements coupled with low-power electronics that reflect, refract, and manipulate electromagnetic waves [33]. These surfaces are equipped with an array of controllable elements that, when dynamically controlled, empower RIS to modify signal propagation characteristics, leading to improved signal strength, coverage extension, interference mitigation, and energy efficiency. This has resulted in several applications of RIS-aided communications in 5G and beyond technologies, IoT, wireless virtual and augmented reality, connected vehicles, and disaster recovery and remote areas coverage [34–37].

8.2.2 Joint Radar-Communication (JRC)

Over recent years, the merge of sensing and signaling has also attracted significant research attention for the development of next-generation systems, to fulfill ever-increasing demands of bandwidth and capacity. Efficient utilization of the RF spectrum is achieved through the shared allocation of spectrum resources for both sensing and signaling purposes, as highlighted in [38]. An optimization-driven approach to beamforming design has been explored to facilitate the simultaneous operation of MIMO radar and cellular systems in the same spectral space, as discussed in [39]. The development of JSAC, such as in the case of the JRC systems, aims to streamline the sharing of both hardware infrastructure and frequency bands, as evidenced in [40–43]. In the existing literature, the integration of communication and sensing-based transmission has mainly been investigated for single antenna systems, as mentioned in [44, 45]. In such systems, a notable performance trade-off arises between the radar and communication functions, as exemplified by the study in [44], which highlights a substantial peak average power ratio along with a constrained dynamic range.

Alternatively, the 802.11ad wireless local area network protocol, as discussed in [46, 47], operates at the 60 GHz band and utilizes a limited number of antennas, primarily supporting short-range sensing applications. In contrast, there are advanced antenna configurations in the context of multiple-input multiple-output (MIMO) systems, as demonstrated in [18, 39, 48], which involve multiple radar users and multiple communication users. These MIMO Joint Radar Communication (JRC) systems offer increased degrees of freedom, resulting in enhanced system performance. To further optimize the performance of MIMO JRC systems, various information embedding techniques have been proposed, as discussed in [49]. In the context of MIMO JRC systems, the design of transmit waveforms plays a critical role in achieving high information transmission rates to designated users while ensuring the reliable operation of the radar system. In [18], the transmit waveform design focuses on minimizing multiuser interference (MUI) to achieve a desired radar beam pattern. Furthermore, [39] proposes precoding solutions to align the transmit signals with the desired radar beam pattern while satisfying communication-related constraints. Additionally, Liu et al. [48] decomposes the precoding matrix into two parts, one dedicated to communication and the other to the radar system. It is worth noting that these techniques are primarily developed for fully digital transceivers, featuring one RF chain per antenna. However, this approach can lead to high power consumption and implementation costs, especially in large-scale systems [39, 48].

Hybrid analog-digital transceivers, optimized for scenarios with a limited number of RF chains, are typically configured using a two-stage beamforming approach. This involves employing a digital beamformer at the baseband stage, followed by an analog beamformer applied in the RF domain. Moreover, [50–53] have explored this design paradigm, often considering hardware simplifications for the conversion units, as indicated in [54]. In instances where systems have only a small number of RF chains, the capacity to support multiuser communication may be limited. Addressing this limitation, the study in [55] introduces a symbol-level precoding approach. This method primarily employs analog components, such as phase shifters (PSs) and VGAs, at the transmitter. Despite its straightforwardness, this nonlinear transmit signal design facilitates the transmission of multiple concurrent data streams, enabling efficient multiuser communication while achieving notable gains in power efficiency and hardware simplicity.

8.2.3 Integration of RIS and JRC

RISs not only enhance spectrum and energy efficiency in wireless communications but also hold promise for advancing ISAC by improving sensing coverage, accuracy, and resolution. Traditionally, ISAC relies on LoS links between transmitters and targets, posing challenges for non-LoS (NLoS) scenarios. RISs offer a potential solution by establishing additional LoS links in NLoS areas, as depicted in Figure 8.1. Moreover, RISs can augment sensing and localization accuracy by providing alternative angles to view targets with existing LoS connections [56, 57]. Effective deployment of RISs requires careful consideration of new sensing requirements and real-time optimization of their reflecting properties, accounting for novel sensing metrics and cochannel interference from sensing and communication tasks. Managing multiple distributed RIS deployments and passive network information poses additional complexities in this regard.

On the flip side, ISAC can reciprocally enhance wireless communication performance involving RISs. A critical technical hurdle in RIS-enabled wireless communications is the challenge of acquiring RIS-related channel state information (CSI), as RIS lacks signal processing capabilities, making tasks

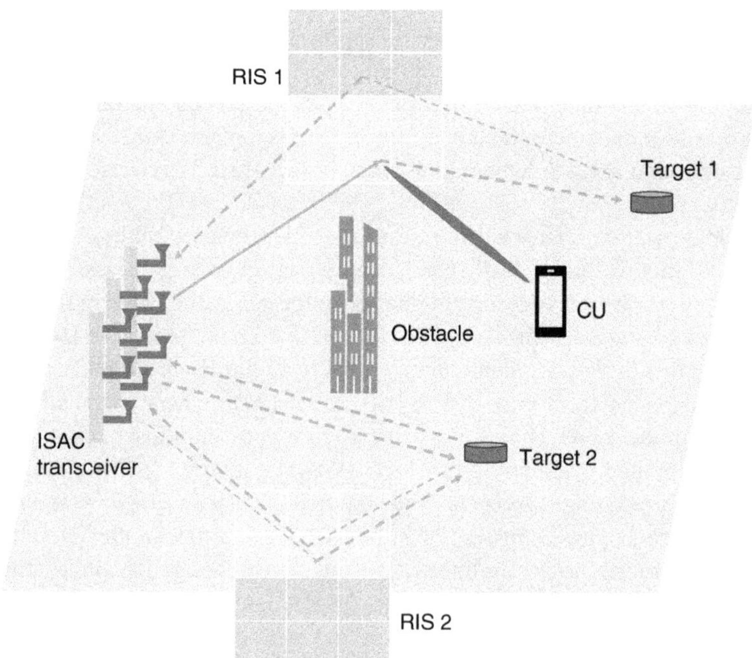

Figure 8.1 RIS assisting ISAC.

like beam tracking and alignment problematic. Here, ISAC can play a crucial role by leveraging its sensing capabilities to measure parameters relevant to communication users (such as angle of arrival/departure), offering an alternative method to obtain CSI that facilitates passive RIS beamforming.

Motivated by the success in both communication and radar fields, the RIS technique has also been introduced for JRC systems, which have the potential to provide extra spatial degrees of freedom to enhance either the transmission or the detection performance of JRC systems [58]. The initial studies carried out in this area have only regarded the RIS as an additional component for communication purposes and the radar sensing has not been considered to be affected by the RIS [59–61]. However, the diversity offered by the RIS has not been utilized completely by such studies and designs. Moreover, there have also been some research works focusing on the JRC system design with a passive RIS, and a semidefinite relaxation (SDR) based design has been presented to overcome the nonconvex radar detection constraint [62]. A double-RIS-aided JRC system has been investigated in [63], where the transceiver is optimized by maximizing the communication signal-to-interference-plus-noise ratio (SINR). Moreover, to unify the radar and communication beamforming designs, the JRC transceiver design has been formulated as a weighted mutual information maximization problem by the author in [64].

8.3 Waveform Design for ISAC Systems

Waveform design is crucial in ISAC systems, focusing on creating a dual-functional waveform capable of both sensing and communication by sharing signaling resources to achieve integration gains. Depending on the level of integration, ISAC waveforms can range from the most loosely coupled approach (time-/frequency-/spatial-/code-division) to the most tightly coupled approach (fully unified waveform). A short review of the ISAC waveform design with nonoverlapping resource allocation, followed by methods for fully unified waveform design is discussed further.

8.3.1 Nonoverlapping Resource Allocation

It is evident that S&C can be scheduled on orthogonal or nonoverlapping wireless resources to prevent interference. This can be achieved across various domains, including temporal, spectral, spatial, or code domains, commonly referred to as time-division, frequency-division, spatial-division, and code-division ISAC, respectively.

Time-division ISAC is the most loosely coupled waveform design, making it easy to integrate into existing commercial systems. In [65], a joint radar-communication waveform design was introduced, where the transmission duration is divided into radar and radio cycles. Specifically, a frequency-modulated continuous waveform (FMCW) with up- and down-chirp modulations is used for radar sensing, while various modulation schemes, such as binary phase shift keying (BPSK), pulse position modulation (PPM), and on-off keying (OOK), can be flexibly employed for communication. Recently, time-division ISAC has been implemented in several commercial wireless standards, such as IEEE 802.11p and IEEE 802.11ad. In these standards, the channel estimation field (CEF) or pilot signals, initially designed for channel estimation, are utilized for radar sensing [46, 47, 66–68].

Frequency-division ISAC is typically built on an orthogonal frequency division multiplexing (OFDM) waveform. While it lacks the flexibility of its time-division counterpart, where any S&C waveforms can be used, it allocates S&C functionalities to different subcarriers based on channel conditions, required KPIs for S&C, and the transmitter's power budget [69, 70]. For example, in [70], a power allocation and subcarrier selection scheme are developed to minimize transmit power while meeting both mutual information and achievable data rate constraints for radar sensing and communications.

Spatial-division ISAC has garnered attention recently due to advancements in MIMO and mMIMO technologies [71]. In these approaches, S&C are carried out over orthogonal spatial resources, such as different antenna groups [39]. When the communication channel is primarily dominated by a LoS component, S&C waveforms can be transmitted over different spatial beams, which exhibit strong orthogonality in the mMIMO regime [72]. Conversely, if the communication channel consists of rich scattering paths, the sensing waveform can be projected into the channel's null space to prevent interference with communication functions [73, 74].

Besides the time, frequency, and spatial division methods, *code-division ISAC* waveform designs are also regarded as viable solutions. In these designs, sensing and communication signals are transmitted using orthogonal or quasi-orthogonal sequences [75, 76]. For instance, a novel code-division ISAC scheme was introduced in [77], which utilized a code-division OFDM ISAC signal along with a corresponding successive interference cancellation-based receiver design. This approach aimed to achieve the benefits of code-division multiplexing (CDM) and enhance the reliability of both sensing and communication.

While the aforementioned designs are easy to implement on hardware platforms, they tend to have poor spectral and energy efficiency. To maximize integration gains, it is advantageous to design a fully unified ISAC waveform, where temporal, spectral, spatial, and signaling resources are shared between sensing and communication [78, 79].

8.3.2 Fully Unified Waveforms

Fully unified ISAC waveforms are typically developed based on three main philosophies: sensing-centric design (SCD), communication-centric design (CCD), and joint design (JD) approaches [40, 80].

SCD seeks to integrate communication functionality into existing sensing waveforms and infrastructures, ensuring that sensing performance is primarily maintained. However, a purely radar-sensing waveform is not suitable for communication as it lacks signaling information. Therefore, the core of SCD involves embedding data into a sensing waveform without significantly compromising its sensing capabilities. Early SCD schemes primarily concentrated on time-frequency domain embedding, using chirp signals-commonly utilized in various radar applications—as carriers of information [81, 82]. Chirp signals feature a large time-bandwidth product and can produce narrow pulses with high peak power from long-duration pulses with low peak power through pulse compression and matched filtering methods, thereby providing excellent range estimation performance. Recent SCD approaches aiming to integrate communication capabilities into MIMO radar systems have explored embedding data into the radar's spatial domain [83–85]. A classical method involves using the sidelobe level of the MIMO radar beampattern to represent each communication symbol, while reserving the main beam exclusively for target sensing [83]. Additionally, to ensure optimal radar sensing performance, a promising SCD approach involves implementing ISAC through index modulation. An advanced study introduced an index modulation scheme using carrier agile phased array radar (CAESAR) to enable ISAC capabilities, specifically termed as the multicarrier agile joint radar-communication (MAJoRCom) system [86, 87]. This system operates by randomly varying carrier frequencies on a pulse-by-pulse basis and assigning these frequencies randomly to each antenna element, thereby introducing agility across both spatial and frequency domains.

It is to be noted that many SCD schemes typically utilize slow-time coding, which involves interpulse modulations rather than within-pulse methods. This approach links the bit rate directly to the pulse repetition frequency (PRF) of the radar. Therefore, while SCD enhances sensing performance, its utility is constrained to scenarios that demand low to moderate data rates. Unlike SCD, *CCD* integrates sensing functionality into an existing communication waveform or system, prioritizing communication as the primary function to ensure. Essentially, any communication waveform can be adapted for monostatic sensing since the transmitter has full knowledge of the waveform. However, the introduction of random communication data can significantly impair the sensing performance of the system. A prominent CCD approach involves utilizing OFDM waveforms for radar sensing, which has gained increased interest recently. This interest stems from their compatibility with modern 4G and 5G standards [44, 88, 89]. The OFDM-based ISAC waveform, while ensuring robust communication performance, faces limitations in its sensing capabilities. Key properties essential for radar systems are often overlooked. Firstly, maintaining a constant envelope is crucial to maximize transmission power without signal distortion, thereby maximizing the signal-to-noise ratio (SNR) of received echo signals. Secondly, addressing clutter interference, critical for effective radar sensing, is rarely prioritized in CCD approaches. Lastly, the sensing waveform must exhibit strong correlation properties to minimize temporal, spectral, and angular sidelobes, thereby reducing the risk of false target detection. Improving the sensing performance of CCD ISAC waveforms requires effective shaping that adheres to these specific sensing constraints.

While SCD and CCD schemes achieve partial integration of ISAC, they struggle to establish a scalable balance between sensing and communication functionalities. These approaches represent two extremes in ISAC waveform design, ensuring either sensing or communication performance while limiting the potential of the other. Addressing this challenge, *JD* emerges as a promising methodology. Unlike SCD and CCD, JD seeks to develop ISAC waveforms from scratch, rather than adapting existing sensing and communication waveforms. This approach introduces additional degrees of freedom and flexibility, thereby enhancing both sensing and communication performance concurrently [18, 39, 48, 90, 91]. The representative examples of the different ISAC waveform designs along with their pros and cons are summarized in Table 8.1.

8.4 Optimal Waveform Design for RIS-Assisted Mimo JRC System: A Case Study

8.4.1 System Model

Let us assume, a MIMO JRC system equipped with a uniform linear array (ULA) of $N_T = N_{T,v} \times N_{T,h}$ antenna elements which simultaneously serves N_R single-antenna downlink user terminals and transmits radar probing waveforms toward the targets. To reduce the hardware complexity, we propose an analog architecture for the MIMO JRC system as shown in Figure 8.1. Such an architecture is implemented by one PS module connected to each transmit antenna before

Table 8.1 Waveform designs for ISAC systems.

Waveform designs	Methodologies	Representative techniques	Pros	Cons
Nonoverlapped resource allocation	Time division	• Target detection (TD) between FMCW and comms modulations [65] • Using CEF of IEEE 802.11ad frame for sensing [46]	Easy to implement	Low efficiency
	Frequency division	• Subcarrier allocation between S&C in OFDM [70]		
	Spatial division	• Antenna separation between S&C [39]		
Fully unified waveform	Sensing centric	• Using chirp signals as information carriers [82] • MIMO radar sidelobe embedding approach [83] • Index modulation [86, 92]	Guaranteed sensing performance	Low information rate/spectral efficiency
	Communication centric	• Using OFDM for sensing [44]	Guaranteed comms performance	Unreliable sensing performance
	Joint design	• Waveform design based on optimization [18, 39, 48, 91]	Scalable S&C performance tradeoff	High complexity

the power amplifiers (PAs). As shown in Figure 8.1, this phase-shifting network is driven by a common VGA, with the local oscillator (LO) circuitry, that tunes the common modulus of the output transmission signals. In this chapter, we consider also an RIS-assisted JRC system that is based on the analog architecture in Figure 8.1 along with an RIS component as shown in Figure 8.4.

8.4.1.1 Communication Model

The received signals of the downlink communication users $\mathbf{Y}$ can be expressed as:

$$\mathbf{Y} = \mathbf{H}\mathbf{X}\mathbf{Q} + \mathbf{Z}, \tag{8.1}$$

where $\mathbf{X} = [\mathbf{x}_1, \dots, \mathbf{x}_M] \in \mathcal{F}^{\mathcal{N}_T \times \mathcal{M}}$ is the transmit signal matrix employed for both radar and communication operations with $\mathcal{F}^{N_T \times M}$ being the set of complex unit-modulus numbers, that is, $\mathcal{F} = \{\omega \in \mathbb{C} \mid |\omega| = \infty\}$ and M is the length of the communication frame/radar pulse. The channel matrix $\mathbf{H} = [\mathbf{h}_1, \dots, \mathbf{h}_{N_R}]^T \in \mathbb{C}^{N_R \times N_T}$ represents the frequency flat Rayleigh fading channel entries between the JRC system and the communication users. The CSI is assumed to be perfectly known at the transceiver. The matrix $\mathbf{Q} = [\mathbf{q}_1, \dots, \mathbf{q}_M]$ is the $M \times M$ diagonal matrix of complex entries that correspond to the common modulus set by the VGA per frame by assuming that the phase components of the diagonal entries of $\mathbf{Q}$ may be absorbed by the PSs, without generality loss. Note that unlike existing literature where the transmit signal matrix is considered for a fully digital system [18], in this chapter the transmit signal $\mathbf{XQ}$ takes into account the impact of low complexity analog architecture through $\mathbf{Q}$ matrix entries. The matrix $\mathbf{Z} = [\mathbf{z}_1, \dots, \mathbf{z}_M] \in \mathbb{C}^{N_R \times M}$ represents the independent and identically distributed complex additive white Gaussian noise where $\mathbf{z}_m \sim \mathcal{CN}(0, \sigma_z^2 \mathbf{I}), \forall m$.

In the presence of an $N_F = N_{F,v} \times N_{F,h}$ RIS, there are additional transmission paths between the BS and the user terminals (UTs), and thus, the UTs' received signals' matrix $\mathbf{Y}$ at M symbol times is given by

$$\mathbf{Y} = (\mathbf{H} + \mathbf{G}_2 \boldsymbol{\Omega} \mathbf{G}_1)\mathbf{XQ} + \mathbf{Z}, \tag{8.2}$$

where the matrices $\mathbf{G}_1 \in \mathbb{C}^{N_F \times N_T}$ and $\mathbf{G}_2 \in \mathbb{C}^{N_R \times N_F}$ contain the coefficients of the channels between the BS and the RIS, and the RIS and the M UTs, respectively and $\boldsymbol{\Omega} \in \mathcal{F}^{N_F \times N_F}$ is the diagonal reflection matrix of the RIS.

The channels are modeled as follows [93, 94]

$$\mathbf{h}_n = \sqrt{\alpha_{0,n}} \left(\sqrt{\frac{K_{0,n}}{K_{0,n}+1}} \bar{\mathbf{h}}_n + \sqrt{\frac{1}{K_{0,n}+1}} \tilde{\mathbf{h}}_n \right), \tag{8.3}$$

$$\mathbf{G}_1 = \sqrt{\alpha_1} \left(\sqrt{\frac{K_1}{K_1+1}} \bar{\mathbf{G}}_1 + \sqrt{\frac{1}{K_1+1}} \tilde{\mathbf{G}}_1 \right), \tag{8.4}$$

$$\mathbf{g}_{2,n} = \sqrt{\alpha_{2,n}} \left(\sqrt{\frac{K_{2,n}}{K_{2,n}+1}} \bar{\mathbf{g}}_{2,n} + \sqrt{\frac{1}{K_{2,n}+1}} \tilde{\mathbf{g}}_{2,n} \right), \tag{8.5}$$

where $\mathbf{h}_n$, $\mathbf{g}_{1,n}$, and $\mathbf{g}_{2,n}^T$ are the column vectors that include the channel coefficients between the BS and the n UT, between the BS and the RIS, and the RIS and the N_R UTs, $\sqrt{\alpha_{0,n}} = d_{0,k}^{-r}$, $\sqrt{\alpha_1} = d_1^{-r}$, $\sqrt{\alpha_{2,n}} = d_{2,n}^{-r}$, $d_{0,n}$, d_1, $d_{2,n}$ are the distances between the BS and the nth UT, the BS and the RIS, the RIS and the nth UT, respectively, r is the path loss coefficient, $K_{0,n}$, K_1, and $K_{0,n}$ are the Rician distribution K-factors, $\bar{\mathbf{h}}_n \in \mathbb{C}^{N_T}$, $\bar{\mathbf{G}}_1 \in \mathbb{C}^{N_F \times N_T}$ and $\bar{\mathbf{G}}_{2,n} \in \mathbb{C}^{N_F}$ are the deterministic line-of-sight (LoS) components, $\tilde{\mathbf{h}}_n \in \mathbb{C}^{N_T}$, $\tilde{\mathbf{G}}_1 \in \mathbb{C}^{N_F \times N_T}$ and $\tilde{\mathbf{G}}_{2,n} \in \mathbb{C}^{N_F}$ are the random scattered/non-LoS components modeled as complex Gaussian variables of zero mean and unit variance.

The deterministic components of the channels are given by

$$\overline{\mathbf{h}}_n = \mathbf{a}(\theta_{0,n}, \phi_{0,n}, N_{T,v}, N_{T,h}), \tag{8.6}$$

$$\overline{\mathbf{G}}_1 = \mathbf{a}^H(\theta_1, \phi_1, N_{F,v}, N_{F,h})\mathbf{a}(\theta_1', \phi_1', N_{T,v}, N_{T,h}), \tag{8.7}$$

$$\overline{\mathbf{g}}_{2,n} = \mathbf{a}(\theta_{2,n}, \phi_{2,n}, N_{F,v}, N_{F,h}), \tag{8.8}$$

where $\theta_{0,n}$, $\phi_{0,n}$, θ_1, ϕ_1, θ_1', ϕ_1', $\theta_{2,n}$, and $\phi_{2,n}$ are the angular parameters, the array response $\mathbf{a}(\theta, \phi, A_v, A_h)$ for a uniform planar array (UPA) with $A = A_v \times A_h$ is defined as:

$$\mathbf{a}(\theta, \phi, A_v, A_h) = \mathbf{a}_v(\theta, A_v) \otimes \mathbf{a}_h(\theta, \phi, A_h), \tag{8.9}$$

$\theta \in (0, \pi)$ / $\phi \in (-\pi/2, \pi/2)$, is the vertical/horizontal angular parameter, respectively, and $\mathbf{a}_v(\theta, A_v)$ and $\mathbf{a}_h(\theta, \phi, A_h)$ are defined as follows:

$$\mathbf{a}_v(\theta, A_v) = \left[1, e^{j2\pi \frac{d}{\lambda} \sin(\theta)}, \dots, e^{j2\pi \frac{d}{\lambda}(A_v-1)\sin(\theta)}\right], \tag{8.10}$$

$$\mathbf{a}_h(\theta, \phi, A_h) = \left[1, e^{j2\pi \frac{d}{\lambda} \cos(\theta)\sin(\phi)}, \dots, e^{j2\pi \frac{d}{\lambda}(A_h-1)\cos(\theta)\sin(\phi)}\right]. \tag{8.11}$$

The random components of the channels are given by

$$\tilde{\mathbf{h}}_n = \mathbf{R}_{h,n}^{1/2}\mathbf{h}_n', \ \tilde{\mathbf{G}}_1 = \mathbf{R}_r^{1/2}\mathbf{G}_1'\mathbf{R}_t^{1/2}, \ \tilde{\mathbf{g}}_{2,n} = \mathbf{R}_{g,n}^{1/2}\mathbf{g}_{2,n}', \tag{8.12}$$

where $\mathbf{h}_n' \in \mathbf{C}^{N_T}$, $\mathbf{G}_1' \in \mathbf{C}^{N_F \times N_T}$, $\mathbf{g}_{2,n}' \in \mathbf{C}^{N_F}$ are complex Gaussian matrices with independent and identically distributed (i.i.d.) entries of zero mean and unit variance; $\mathbf{R}_{h,n} \in \mathbf{C}^{N_T \times N_T}$ ($\mathbf{R}_{g,n} \in \mathbf{C}^{N_F \times N_F}$) are the correlation matrices between the nth UT and the BS (RIS) and $\mathbf{R}_t \in \mathbf{C}^{N_T \times N_T}$ ($\mathbf{R}_r \in \mathbf{C}^{N_F \times N_F}$) is the RIS transmit (receive) correlation matrix. The aforementioned correlation matrices can be approximated for the considered UPA-based BS and RIS systems as [95]

$$\mathbf{R}_{h,n} \approx \mathbf{R}_{h,n,x} \otimes \mathbf{R}_{h,n,y}, \tag{8.13}$$

$$\mathbf{R}_t \approx \mathbf{R}_{t,x} \otimes \mathbf{R}_{t,y}, \tag{8.14}$$

$$\mathbf{R}_r \approx \mathbf{R}_{r,x} \otimes \mathbf{R}_{r,y}, \tag{8.15}$$

$$\mathbf{R}_{g,n} \approx \mathbf{R}_{g,n,x} \otimes \mathbf{R}_{g,n,y}, \tag{8.16}$$

where $\mathbf{R}_{h,n,x}$, $\mathbf{R}_{t,x}$, $\mathbf{R}_{r,x}$, $\mathbf{R}_{g,n,x}$ and $\mathbf{R}_{h,n,y}$, $\mathbf{R}_{t,y}$, $\mathbf{R}_{r,y}$, $\mathbf{R}_{g,n,y}$ are the corresponding correlation matrices of the horizontal and vertical domains of the rectangular RIS, respectively. For such a correlation matrix (either on the horizontal or on the vertical domain) $\mathbf{R}$, we adopt the well-known exponential model [37] according to which the (i, j)th entry of $\mathbf{R}$, $r_{i,j}$, is given by

$$r_{i,j} = \begin{cases} \rho e^{j\theta|i-j|}, & i \geq j, \\ \rho e^{-j\theta|i-j|}, & i < j, \end{cases} \tag{8.17}$$

where $0 \leq \rho < 1$ and θ is the corresponding horizontal or vertical angular parameter. It is assumed that the channel statistics, that is, the LoS components (8.6)–(8.8) and the correlation matrices (8.13)–(8.16) vary quite slowly compared to the NLoS components that are rapidly varying.

The communication part of the system aims at the transmission of a desired symbol $s(n,m) \in \mathcal{O}, 1 \leq n \leq N_R, 1 \leq m \leq M$, from the BS to the nth UT in the mth symbol time. $\mathcal{O}$ is the set of the employed constellation points. In each symbol time, the vector of the symbols to be transmitted to the UTs is denoted by $\mathbf{s}_m \in \mathcal{O}^{\leq N_R \times 1}$. Given the CSI and $\mathbf{S} = [\mathbf{s}_1, \dots, \mathbf{s}_N]$, that includes the symbols to be transmitted to the M UTs at a block of N symbol times, the BS aims at the design of a transmit signal matrices, $\mathbf{X}$ and $\mathbf{Q}$ and the reflection matrix $\mathbf{\Omega}$ for the case of the RIS-assisted system, such that matrix of the received signals at the UTs during these N symbol times $\mathbf{Y}$ is as close to $\mathbf{S}$, as possible.

The latter objective may be achieved by minimizing the so-called "MUI energy" [96], given by

$$f(\mathbf{X},\mathbf{Q}) = \|\mathbf{HXQ} - \mathbf{S}\|_F^2, \tag{8.18}$$

$$f'(\mathbf{\Omega},\mathbf{X},\mathbf{Q}) = \|(\mathbf{H} + \mathbf{G}_2\mathbf{\Omega}\mathbf{G}_1)\mathbf{XQ} - \mathbf{S}\|_F^2, \tag{8.19}$$

for non-RIS-assisted and RIS-assisted systems, respectively. Observe that the received signal of the nth UT in the mth symbol time can be written for both cases as:

$$y_{n,m} = s_{n,m} + \underbrace{\mathbf{h}_n^T\mathbf{x}_m - s_{n,m}}_{\text{MUI at the } n\text{th user}} + z_{n,m}, \tag{8.20}$$

$$y_{n,m} = s_{n,m} + \underbrace{(\mathbf{h}_n^T + \mathbf{g}_{2,n}^T\mathbf{\Omega}\mathbf{g}_{1,n})\mathbf{x}_m - s_{n,m}}_{\text{MUI at the } n\text{th user}} + z_{n,m}. \tag{8.21}$$

Then, according to (8.21), the receive SINR per block of M symbols for the nth UT, is defined for both cases as:

$$\xi_n = \frac{\mathbb{E}\{|s_{n,m}|^2\}}{\mathbb{E}\{|\mathbf{h}_n^T\mathbf{x}_m - s_{n,m}|^2\} + N_0}, \tag{8.22}$$

$$\xi_n = \frac{\mathbb{E}\{|s_{n,m}|^2\}}{\mathbb{E}\{|(\mathbf{h}_n^T + \mathbf{g}_{2,n}^T\mathbf{\Omega}\mathbf{g}_{1,n})\mathbf{x}_m - s_{n,m}|^2\} + N_0}, \tag{8.23}$$

where $s_{n,m}$ is the (n,m)th entry of the symbol matrix $\mathbf{S}$. In (8.22) and (8.23), the expectation operator in the numerators is applied to $s_{n,m}$ and in the denominators is applied to $s_{n,m}$ and $\mathbf{x}_m$, as well.

It has been shown in [96] that an approximation of the achievable information rate for the nth user is a function of ξ_n in (8.22) and (8.23). Thus, the achievable sum rate of the communication part can be approximated by

$$r = \sum_{n=1}^{N_R} \log_2(1 + \xi_n), \tag{8.24}$$

where ξ_n is calculated by (8.22) or (8.23) depending on the considered regime (non-RIS-assisted or RIS-assisted).

Let us assume that in (1) the symbols $s_{m,n}$, $1 \leq m \leq M$, $1 \leq n \leq N$ are drawn from the same constellation set $\mathcal{O}$ and have fixed energy. Then, the signal power $\mathbb{E}\{|s_{m,n}|^2\}$ in (8.22) and (8.23) is also fixed and thus, the SINR expression per user can be maximized by minimizing the MUI energy in (8.18) and (8.19). From (8.24), it is straightforward to see that by maximizing the SINR of the UTs, their achievable rate is maximized, as well. In other words, the MUI energy minimization is a suitable criterion for optimizing the rate performance of the system.

In the proposed design, we aim at the optimization of the communication part of the JRC system through the minimization of the MUI energy subject to the constraints related to the radar waveform design and additionally due to the employed hardware, that is, unit-modulus signals due to the PSs at the BS and the RIS (whenever available) and maximum transmission power constraints due to the VGA/PAs at the BS, again.

8.4.1.2 Radar Model

The transmit beampattern for the JRC system that points to the targets of interest can be expressed as:

$$P_T(\theta) = \mathbf{a}^H(\theta, \phi, N_{T,v}, N_{T,u})\mathbf{R}_o\mathbf{a}_T(\theta, \phi, N_{T,v}, N_{T,u}), \tag{8.25}$$

where $\mathbf{a}_T(\theta, \phi, N_{T,v}, N_{T,u})$ is defined as in (8.9)–(8.11). The covariance matrix associated with the transmit signal matrix $\mathbf{XQ}$ which takes into account the analog architecture through $\mathbf{Q}$, is

$$\mathbf{R}_o = \frac{1}{M}\mathbf{XQQ}^H\mathbf{X}^H, \tag{8.26}$$

where using the assumption $M \geq N_T$ without loss of generality, $\mathbf{R}_o$ in (8.26) is positive definite. Following [97], designing the transmit covariance matrix $\mathbf{R}_o$ is equivalent to designing the transmit radar beam pattern in (8.25) for uncorrelated waveforms where the MIMO radar exhibits a high degree of freedom (DoF) in comparison to the conventional phased array radar. Having described the desired cost function and the desired features of the radar waveform, the optimization problem via which the transmission signals are derived for the low-complexity analog architecture are defined next.

8.4.2 Optimal Waveform Design for Non-RIS-Assisted System

This section presents the definition of the optimization problems out of which the optimal transmit waveform and the corresponding algorithmic solution for the non-RIS-assisted system are proposed and derived [95]. We consider $\mathbf{R}_D$ as the Hermitian positive semidefinite covariance matrix corresponding to a well-designed radar beampattern [18]. The transmit waveform problem design for the JRC system involves the derivation of the optimal $\mathbf{X}$ and $\mathbf{Q}$ matrices per block of M symbols through minimizing the MUI function in (8.18), that is

$$(\mathcal{P}_1): \quad \min_{\mathbf{X},\mathbf{Q}} \frac{1}{2} \|\mathbf{S} - \mathbf{HXQ}\|_F^2$$

$$\text{s.t. } \mathbf{X} \in \mathcal{F}^{N_T \times M} \tag{8.27}$$

$$\frac{1}{M}\mathbf{XQQ}^H\mathbf{X}^H = \mathbf{R}_D \tag{8.28}$$

$$|\text{diag}\{\mathbf{Q}\}|^2 \preceq P_{\max}\mathbf{1}, \tag{8.29}$$

where $\mathbf{1}$ is the vector of ones and the constraint in (8.27) is due to the unit-modulus entries in $\mathbf{X}$, (8.28) imposes the desired beampattern to the transmit waveform and (8.29) is a maximum power budget-related constraint ($P_{\max}$).

Problem ($\mathcal{P}_1$) is nonconvex due to the nonconvex objective function and non-convex constraint set. It is in general, a difficult NP-hard problem to solve [98]. In the following, a solution is developed based on the Alternating direction method of multipliers (ADMM). It is a primal-dual method that applies an alternating minimization approach to efficiently solve constrained optimization problems. ADMM-based solutions have been successfully applied to tackle problems that involve nonconvex objective functions and constraints, similar to the ones of $\mathcal{P}_1$ [99–102]. For convenience, using the auxiliary matrix $\mathbf{Z} \in \mathbb{C}^{N_T \times M}$, the problem ($\mathcal{P}_1$) is rewritten as [95]

$$(\mathcal{P}_2): \quad \min_{\mathbf{Z},\mathbf{X},\mathbf{Q}} \frac{1}{2} \|\mathbf{S} - \mathbf{HZ}\|_F^2$$

$$\text{s.t. } \mathbf{Z} = \mathbf{XQ} \tag{8.30}$$

$$\mathbf{X} \in \mathcal{F}^{N_T \times M} \tag{8.31}$$

$$\frac{1}{M}\mathbf{ZZ}^H = \mathbf{R}_D \tag{8.32}$$

$$|\text{diag}\{\mathbf{Q}\}|^2 \preceq P_{\max}\mathbf{1}. \tag{8.33}$$

Following that, the augmented Lagrangian function of the optimization problem can be written as:

$$\mathcal{L}_1(\mathbf{Z}, \mathbf{X}, \mathbf{Q}, \mathbf{\Lambda}) = \frac{1}{2} \|\mathbf{S} - \mathbf{HZ}\|_F^2 + \text{tr}\left(\mathbf{\Lambda}^H(\mathbf{Z} - \mathbf{XQ})\right) + \frac{\rho}{2} \|\mathbf{Z} - \mathbf{XQ}\|_F^2, \tag{8.34}$$

where $\rho \in \mathbb{R}_+$ is the hyperparameter, which is a scalar quantity and $\mathbf{\Lambda} \in \mathbb{C}^{N_T \times M}$ is the Lagrange multiplier matrix.

Then, the application of the ADMM for solving $\mathcal{P}_2$ involves the alternating minimization steps, given by [95]

$$(\mathcal{P}_{2A}) : \mathbf{Z}_{k+1} = \arg\min_{\mathbf{Z} \in \mathcal{Z}} \mathcal{L}_1(\mathbf{Z}, \mathbf{X}_k, \mathbf{Q}_k, \mathbf{\Lambda}_k), \tag{8.35}$$

$$(\mathcal{P}_{2B}) : \mathbf{X}_{k+1} = \arg\min_{\mathbf{X} \in \mathcal{F}^{N_T \times M}} \mathcal{L}_1(\mathbf{Z}_{k+1}, \mathbf{X}, \mathbf{Q}_k, \mathbf{\Lambda}_k), \tag{8.36}$$

$$(\mathcal{P}_{2C}) : \mathbf{Q}_{k+1} = \arg\min_{\mathbf{Q} \in \mathcal{Q}} \mathcal{L}_1(\mathbf{Z}_{k+1}, \mathbf{X}_{k+1}, \mathbf{Q}, \mathbf{\Lambda}_k), \tag{8.37}$$

where $\mathbf{Z}_k$, $\mathbf{X}_k$, $\mathbf{Q}_k$, and $\mathbf{\Lambda}_k$ are the variable values at the kth iteration of the ADMM sequence and $\mathcal{Z}$ and $\mathcal{Q}$ are the sets associated with the constraints in (8.32) and (8.33), respectively. The dual variables in $\mathbf{\Lambda}_{k+1}$, $1 \leq m \leq M$ are updated via a gradient ascent step [95], that is,

$$\mathbf{\Lambda}_{k+1} = \mathbf{\Lambda}_k + \rho\left(\mathbf{Z}_{k+1} - \mathbf{X}_{k+1}\mathbf{Q}_{k+1}\right). \tag{8.38}$$

We move forward with the derivation to solution in $(\mathcal{P}_{2A})$, that is,

$$\begin{aligned} \min_{\mathbf{Z}} \; & -\text{tr}(\text{re}(\mathbf{S}^H\mathbf{HZ})) + \frac{1}{2} \|\mathbf{HZ}\|_F^2 + \text{tr}(\text{re}(\mathbf{\Lambda}_k^H\mathbf{Z})) \\ & + \frac{\rho}{2}[\|\mathbf{Z}\|_F^2 - 2\text{tr}(\text{re}(\mathbf{Z}^H\mathbf{X}_k\mathbf{Q}_k))] \\ & \text{subject to } \frac{1}{M}\mathbf{ZZ}^H = \mathbf{R}_D. \end{aligned} \tag{8.39}$$

The simplified objective function can be written as, $-\text{tr}\{\text{re}(\mathbf{Z}^H(\mathbf{H}^H\mathbf{S} - \mathbf{\Lambda}_k + \rho\mathbf{X}_k\mathbf{Q}_k))\}$. The problem can be further formulated as [95]:

$$\begin{aligned} \min_{\mathbf{Z}} \; & \left\|\mathbf{Z} - \hat{\mathbf{S}}\right\|_F^2 \\ & \text{subject to } \frac{1}{M}\mathbf{ZZ}^H = \mathbf{R}_D, \end{aligned} \tag{8.40}$$

where $\hat{\mathbf{S}} = \mathbf{H}^H\mathbf{S} - \mathbf{\Lambda}_k + \rho\mathbf{X}_k\mathbf{Q}_k$. The related constraint can be further expressed as, $\frac{1}{M}\mathbf{F}^{-H}\mathbf{ZZ}^H\mathbf{F}^{-1} = \mathbf{I}$, where $\mathbf{F}$ is calculated by the Cholesky decomposition of the covariance matrix $\mathbf{R}_D$, that is, $\mathbf{R}_D = \mathbf{FF}^H$, where the matrix $\mathbf{F}$ is a $N_T \times N_T$ lower triangular matrix. Matrix $\mathbf{F}$ is invertible since the covariance matrix $\mathbf{R}_D$ can be assumed to be positive definite. We let $\tilde{\mathbf{Z}} = \sqrt{\frac{1}{M}}\mathbf{F}^{-1}\mathbf{Z}$ and thus we have

$$\begin{aligned} &\min_{\mathbf{Z}} \quad \left\| \sqrt{M}\mathbf{F}\tilde{\mathbf{Z}} - \hat{\mathbf{S}} \right\|_F^2 \\ &\quad\text{subject to } \tilde{\mathbf{Z}}\tilde{\mathbf{Z}}^H = \mathbf{I}. \end{aligned} \tag{8.41}$$

The above problem in (8.41) represents an orthogonal Procrustes problem (OPP) which admits the following closed-form solution [95]:

$$\tilde{\mathbf{Z}} = \tilde{\mathbf{U}}\mathbf{I}_{N_R\times M}\tilde{\mathbf{V}}^H, \tag{8.42}$$

where $\tilde{\mathbf{U}}\tilde{\mathbf{\Sigma}}\tilde{\mathbf{V}}^H = \mathbf{F}^H\hat{\mathbf{S}}$ represents the singular value decomposition (SVD) of $\mathbf{F}^H\hat{\mathbf{S}}$ where $\tilde{\mathbf{U}}$ and $\tilde{\mathbf{V}}$ are the left and right singular matrices of SVD, respectively, and $\tilde{\mathbf{\Sigma}}$ denotes the diagonal matrix of singular values. Thus, the update for variable $\mathbf{Z}$ is given by

$$\sqrt{\frac{1}{M}}\mathbf{F}^{-1}\mathbf{Z}_{k+1} = \tilde{\mathbf{U}}\mathbf{I}_{N_R\times M}\tilde{\mathbf{V}}^H \Longrightarrow \mathbf{Z}_{k+1} = \sqrt{M}\mathbf{F}\tilde{\mathbf{U}}\mathbf{I}_{N_R\times M}\tilde{\mathbf{V}}^H. \tag{8.43}$$

We now derive the solution to $\mathcal{P}_{2B}$. We have,

$$\begin{aligned} &\min_{\mathbf{X}} \; -\text{tr}(\text{re}(\mathbf{\Lambda}_k^H\mathbf{X}\mathbf{Q}_k)) + \frac{\rho}{2}[-2\text{tr}(\text{re}(\mathbf{Z}_{k+1}^H\mathbf{X}\mathbf{Q}_k)) + \|\mathbf{X}_k\mathbf{Q}_k\|_F^2] \\ &\quad\text{subject to } \mathbf{X} \in \mathcal{F}^{N_T\times M}, \end{aligned} \tag{8.44}$$

The solution to this problem does not admit a closed form and thus, we approximate it by solving the unconstrained problem and then projecting onto set $\mathcal{F}^{N_T\times M}$. That is,

$$\mathbf{X}_{k+1} = \Pi_{\mathcal{F}}\left\{\frac{1}{\rho}\mathbf{Q}_k^{-H}\left(\mathbf{\Lambda}_k^H - \rho\mathbf{Z}_{k+1}^H\right)\right\}, \tag{8.45}$$

where $\Pi_{\mathcal{F}}\{\cdot\}$ is the element-wise projection operator onto set $\mathcal{F}$ and $\mathbf{Q}_k^{-H}$ is the inverse Hermitian transpose of $\mathbf{Q}_k$. The projection operator $\Pi_{\mathcal{F}}\{\cdot\}$ can be computed by solving the following optimization problem [95],

$$\begin{aligned} (\mathcal{P}_3): \quad &\min_{\mathbf{A}_{\mathcal{F}}} \|\mathbf{A}_{\mathcal{F}} - \mathbf{A}\|_F^2 \\ &\text{s.t. } \mathbf{A}_{\mathcal{F}} \in \mathcal{F}, \end{aligned} \tag{8.46}$$

where $\mathbf{A}$ is an arbitrary matrix and $\mathbf{A}_{\mathcal{F}}$ is its element-wise projection of the onto the set $\mathcal{F}$. It can be shown that $(\mathcal{P}_3)$ admits the closed form solution given by

$$\mathbf{A}_{\mathcal{F}}(i_r, i_c) = \begin{cases} 0, & \mathbf{A}(i_r, i_c) = 0 \\ \frac{\mathbf{A}(i_r,i_c)}{|\mathbf{A}(i_r,i_c)|}, & \mathbf{A}(i_r, i_c) \neq 0, \end{cases} \tag{8.47}$$

where $\mathbf{A}_{\mathcal{F}}(i_r, i_c)$ and $\mathbf{A}(i_r, i_c)$ are the elements at the i_rth row – i_cth column of matrices $\mathbf{A}_{\mathcal{F}}$ and $\mathbf{A}$, respectively. The derived inexact solution appears to behave remarkably well as it is verified by the numerical results. Recent literature papers have presented similar results when applying ADMM-based solutions with inexact updates to solve optimization problems [104, 105]. This can be further explained by theoretical results showing that under certain conditions, ADMM with alternating minimization steps that are not carried out exactly is possible to converge [105]. Nevertheless, a detailed theoretical study is beyond the scope of the present work.

Next, we derive the solution to $\mathcal{P}_{2C}$,

$$\begin{aligned} &\min_{\mathbf{Q}} \; -\operatorname{tr}(\operatorname{re}(\boldsymbol{\Lambda}^H\mathbf{XQ})) + \frac{\rho}{2}[-2\operatorname{tr}(\operatorname{re}(\mathbf{Z}^H\mathbf{XQ})) + \|\mathbf{XQ}\|_F^2] \\ &\qquad \text{subject to } |\operatorname{diag}\{\mathbf{Q}\}|^2 \preceq P_{\max}\mathbf{1}. \end{aligned} \tag{8.48}$$

$\mathcal{P}_{2C}$ does not admit a closed-form solution and thus we resort once more to an inexact solution derived by first solving the unconstrained problem and then projecting onto the feasible solution set, that is,

$$\mathbf{Q}_{k+1} = \Pi_Q\left\{\operatorname{diag}\left\{\frac{1}{\rho}\left(\boldsymbol{\Lambda}_k^H - \rho\mathbf{Z}_{k+1}^H\right)\mathbf{X}_{k+1}\left(\mathbf{X}_{k+1}\mathbf{X}_{k+1}^H\right)^{-1}\right\}\right\} \tag{8.49}$$

where the projection operator Π_Q can be derived by solving the following optimization problem [95],

$$\begin{aligned} (\mathcal{P}_4): \quad &\min_{\mathbf{A}_Q} \|\mathbf{A}_Q - \mathbf{A}\|_F^2 \\ &\text{s.t. } \mathbf{A}_Q \in \mathcal{Q}, \end{aligned} \tag{8.50}$$

where $\mathbf{A}$ is again, an arbitrary matrix and $\mathbf{A}_Q$ is its element-wise projection of the onto the set $\mathcal{Q}$. It can be shown that $(\mathcal{P}_4)$ admits the closed form solution given by

$$\mathbf{A}_Q(i,i) = \begin{cases} \mathbf{A}(i,i), & |\mathbf{A}(i,i)|^2 \le P_{\max}, \\ \sqrt{P_{\max}}\frac{\mathbf{A}(i,i)}{|\mathbf{A}(i,i)|}, & |\mathbf{A}(i,i)|^2 > P_{\max}, \end{cases} \tag{8.51}$$

where $\mathbf{A}_Q(i,i)$ and $\mathbf{A}(i,i)$ are the ith diagonal elements of matrices $\mathbf{A}_Q$ and $\mathbf{A}$, respectively.

The alternating optimization procedure (8.35)–(8.38) via which the values of $\mathbf{Z}, \mathbf{X}, \mathbf{Q}$ and $\boldsymbol{\Lambda}$ are derived runs in an iterative manner until convergence is achieved. As for the termination criteria for the alternating minimization procedure, we consider those given by [95],

$$\|\mathbf{Z}_{k+1} - \mathbf{Z}_k\|_F \le \epsilon_z \text{ and} \tag{8.52}$$

$$\|\mathbf{Z}_{k+1} - \mathbf{X}_{k+1}\mathbf{Q}_{k+1}\|_F \le \epsilon_p, \tag{8.53}$$

where ϵ_z and ϵ_p are the corresponding tolerances. The first termination criterion guarantees the convergence of variable $\mathbf{z}_k$ and further the convergence of $(\mathcal{P}_{2A})$. The second one guarantees that the primal feasibility condition of $(\mathcal{P}_2)$ is satisfied. Due to the nonconvex nature of the latter problem, it is also useful to add a termination criterion related to the maximum permitted number of iterations of the ADMM sequence, which here is defined as $K_{\max}$. The complete procedure is summarized in Algorithm 8.1. The computational complexity per iteration is $O(M^3)$ due to the singular value decomposition (SVD) calculation in the solution of $\mathcal{P}_{2A}$ and under the reasonable assumption that in general $M > N_T$.

Algorithm 8.1 Method for non-RIS-assisted JRC system's waveform design [95]

Initialize: $\mathbf{X}_0 \in \mathcal{F}^{N_T \times M}$, $\mathbf{Z}_0$ to be matrix satisfying $\frac{1}{M}\mathbf{Z}_0\mathbf{Z}_0^H = \mathbf{R}_D$ and initial $\mathbf{\Lambda}_0$ with zeros
while The criteria in (8.52) and (8.53) **do**
  $k \leftarrow k+1$
  Calculate $\hat{\mathbf{S}} = \mathbf{H}^H\mathbf{S} - \mathbf{\Lambda}_k + \rho\mathbf{X}_k\mathbf{Q}_k$
  Compute the SVD of $\mathbf{F}^H\hat{\mathbf{S}}$ and calculate $\tilde{\mathbf{Z}}$ by (8.42)
  Update auxiliary matrix $\mathbf{Z}_{k+1}$ using solution (8.43)
  Update waveform matrix $\mathbf{X}_{k+1}$ using solution (8.45)
  Update VGA matrix $\mathbf{Q}_{k+1}$ using solution (8.49)
  Update Lagrange multiplier matrix $\mathbf{\Lambda}_{k+1}$ using (8.38)
end while
return $\mathbf{X}_\star$,$\mathbf{Q}_\star$

We may now comment on the convergence properties of the proposed approach. In literature so far, strong convergence results for ADMM-based solutions have been derived for convex problems having two blocks of variables. Furthermore, for nonconvex problems, strong convergence guarantees are in general unknown and also an open research problem. Problem ($\mathcal{P}_2$) involves three blocks of variables and it involves a nonconvex objective function and feasible solution set. Therefore, the derivation of strong convergence results is an intractable task and beyond the scope of the present work.

8.4.2.1 Simulation Results

In order to evaluate the performance of the proposed method, we set the following simulation parameters: $N_T = 16$, $N_R = 6$, the communication frame/radar pulse length $M = 30$ and maximum power $P_{\max} = 1$ W. We assume a ULA-based setup where antenna elements are spaced by half-wavelength distance, that is, $d = \lambda/2$ where λ can be based on a standard frequency value [18]. We set the number of targets to be three and the corresponding angular target locations at $[-\pi/3, 0, \pi/3]$. The desired $\mathbf{R}_D$ matrix variable is obtained by a least-squares approach such as in [97]. The entries of symbol matrix $\mathbf{S}$ are drawn from a quadrature phase shift keying (QPSK) constellation. The SNR is defined as $1/\sigma_z^2$ and channel matrix $\mathbf{H}$ has complex Gaussian elements.

For comparison with the proposed primal-dual method, we consider the following fully digital baselines:

(a) The state-of-the-art *directional-strict* approach which aims to achieve radar beam pattern with strict equality constraints described in [18].
(b) The *directional trade-off* design which allows a tolerable mismatch between the desired radar beam pattern and the designed one in order to form a balance with the communication performance. The value of the weighting factor for communication and radar operations in the trade-off case is chosen as 0.5;
(c) The *zero MUI* case represents the performance lower bound for MUI and achieves the maximum rate performance.

Figure 8.2 shows the average achievable sum-rate performance described in (8.24) with respect to (w.r.t.) SNR for $N_T = 16$ and $N_R = 6$. It can be observed that the proposed method achieves better rate performance than the existing state-of-the-art directional-strict approach. This is the case since in the proposed approach a relaxed version of the original problem is treated via the optimization of the augmented Lagrangian. That is, the desired beampattern is imposed in the sense that $\rho\|\mathbf{XQ} - \mathbf{Z}\|_F \to 0$. In general, for small ρ values more weight is given to the MUI term compared to the beampattern imposing term. This explains the observed better performance of the proposed approach compared to one of the directional-strict solutions that impose the exact beam pattern to the desired signals. For example, at 10 dB SNR, the proposed method outperforms the directional-strict case by ≈5 bits/s/Hz. Furthermore, the proposed primal-dual approach performs similarly to the trade-off method and the performance in comparison with the zero MUI case is also satisfactory given the simplicity of the hardware architecture that implements the proposed approach.

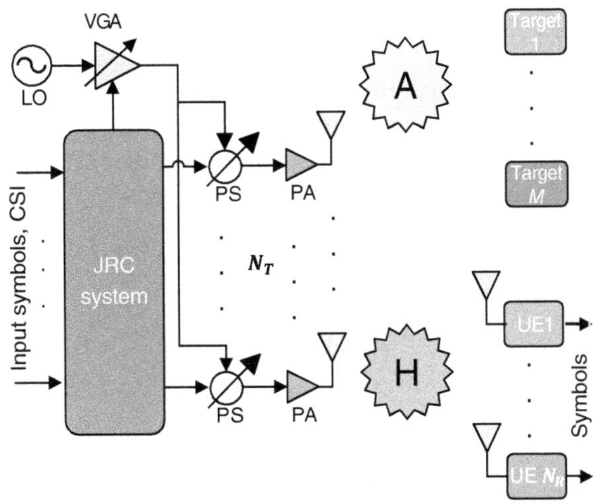

Figure 8.2 A MIMO JRC system with low complexity analog architecture.

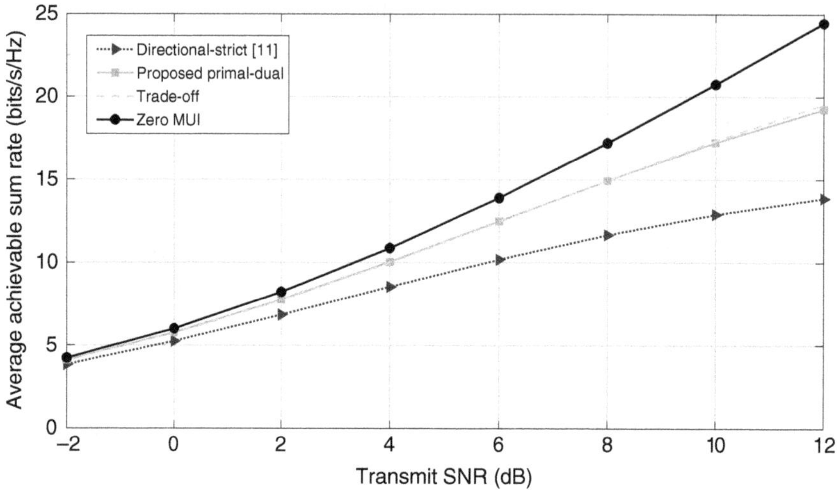

Figure 8.3 Sum rate w.r.t. SNR communication performance comparison for different methods with $N_T = 16, N_R = 6$.

Figure 8.3 shows the radar beam pattern performance plot for the proposed primal-dual method and the baselines for a 16×6 MIMO radar-based JRC system. It can be observed that the proposed primal-dual method exhibits favorable transmit beampattern performance better or similar to the directional-strict baseline at different target locations. In the beampattern plot, the three major lobes for three targets can be observed at the angle locations of $[-\pi/3, 0, \pi/3]$ (Figure 8.4).

8.4.3 Optimal Waveform Design for RIS-Assisted System

This section presents the derivation of the optimal transmit waveform and the corresponding proposed algorithmic solution for the RIS-assisted system. In a similar manner to the non-RIS-assisted system, the MUI minimization problem for the RIS-assisted JRC system can be expressed as:

$$
\begin{aligned}
(\mathcal{P}_5): \quad & \min_{\mathbf{X},\mathbf{Q},\boldsymbol{\Omega}} \frac{1}{2}\|(\mathbf{H} + \mathbf{G}_2\boldsymbol{\Omega}\mathbf{G}_1)\mathbf{X}\mathbf{Q} - \mathbf{S}\|_F^2 \\
& \text{s.t. } \mathbf{X} \in \mathcal{F}^{N_T \times M} \\
& \quad \boldsymbol{\Omega} \in \mathcal{F}_D{}^{N_F \times N_F} \\
& \quad \frac{1}{M}\mathbf{X}\mathbf{Q}\mathbf{Q}^H\mathbf{X}^H = \mathbf{R}_D \\
& \quad |\mathrm{diag}\{\mathbf{Q}\}|^2 \preceq P_{\max}\mathbf{1},
\end{aligned}
$$

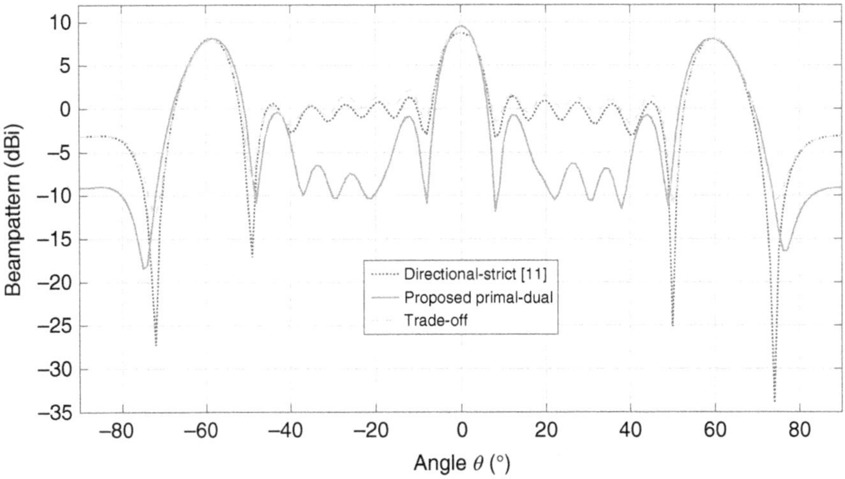

Figure 8.4 Radar beampattern performance for different methods with $N_T = 16$, $N_R = 6$.

where ${\mathcal{F}_D}^{N_F \times N_F}$ is the set of $N_F \times N_F$ diagonal matrices having unit-modulus entries in their main diagonal.

Problem $(\mathcal{P}_5)$ is also nonconvex due to the nonconvex objective function and nonconvex solution set. It is in general a more difficult problem than $\mathcal{P}_1$ due to the optimization of the additional matrix variable $\mathbf{\Omega}$ that lies in the nonconvex set of unit-modulus numbers. We once more resort to the ADMM in order to develop a solution for $(\mathcal{P}_5)$. To that end, we use an auxiliary matrix variable $\mathbf{Z} \in \mathbb{C}^{N_T \times M}$ and rewrite $\mathcal{P}_5$ in the following equivalent form:

$$
\begin{aligned}
(\mathcal{P}_6): \quad & \min_{\mathbf{Z},\mathbf{X},\mathbf{Q},\mathbf{\Omega}} \frac{1}{2} \|\mathbf{S} - (\mathbf{H} + \mathbf{G}_2 \mathbf{\Omega} \mathbf{G}_1)\mathbf{Z}\|_F^2 \\
& \text{s.t. } \mathbf{Z} = \mathbf{X}\mathbf{Q} \\
& \quad \mathbf{X} \in \mathcal{F}^{N_T \times M} \\
& \quad \mathbf{\Omega} \in {\mathcal{F}_D}^{N_F \times N_F} \\
& \quad \frac{1}{M} \mathbf{Z}\mathbf{Z}^H = \mathbf{R}_D \\
& \quad |\text{diag}\{\mathbf{Q}\}|^2 \leq P_{\max} \mathbf{1}.
\end{aligned}
$$

Following that, the augmented Lagrangian function of the optimization problem can be written as:

$$
\begin{aligned}
\mathcal{L}_2(\mathbf{Z}, \mathbf{X}, \mathbf{Q}, \mathbf{\Omega}, \mathbf{\Psi}) = & \frac{1}{2} \|\mathbf{S} - (\mathbf{H} + \mathbf{G}_2 \mathbf{\Omega} \mathbf{G}_1)\mathbf{Z}\|_F^2 + \text{tr}\left(\mathbf{\Psi}^H (\mathbf{Z} - \mathbf{X}\mathbf{Q})\right) \\
& + \frac{\gamma}{2} \|\mathbf{Z} - \mathbf{X}\mathbf{Q}\|_F^2,
\end{aligned} \tag{8.54}
$$

where $\gamma \in \mathbb{R}_+$ is the hyperparameter which is a scalar quantity and $\boldsymbol{\Psi} \in \mathbb{C}^{N_T \times M}$ is the Lagrange Multiplier matrix.

Then, the alternating minimization steps of the ADMM for solving $\mathcal{P}_6$ are given by

$$(\mathcal{P}_{6A}) : \mathbf{Z}_{k+1} = \arg\min_{\mathbf{Z}\in\mathcal{Z}} \mathcal{L}_2(\mathbf{Z}, \mathbf{X}_k, \mathbf{Q}_k, \boldsymbol{\Omega}_k, \boldsymbol{\Psi}_k), \tag{8.55}$$

$$(\mathcal{P}_{6B}) : \mathbf{X}_{k+1} = \arg\min_{\mathbf{X}\in\mathcal{F}^{N_T\times M}} \mathcal{L}_2(\mathbf{Z}_{k+1}, \mathbf{X}, \mathbf{Q}_k, \boldsymbol{\Omega}_k \boldsymbol{\Psi}_k), \tag{8.56}$$

$$(\mathcal{P}_{6C}) : \mathbf{Q}_{k+1} = \arg\min_{\mathbf{Q}\in\mathcal{Q}} \mathcal{L}_2(\mathbf{Z}_{k+1}, \mathbf{X}_{k+1}, \mathbf{Q}, \boldsymbol{\Omega}_k, \boldsymbol{\Psi}_k), \tag{8.57}$$

$$(\mathcal{P}_{6D}) : \boldsymbol{\Omega}_{k+1} = \arg\min_{\boldsymbol{\Omega}\in\mathcal{F}_D^{N_F\times N_F}} \mathcal{L}_2(\mathbf{Z}_{k+1}, \mathbf{X}_{k+1}, \mathbf{Q}_{k+1}, \boldsymbol{\Omega}, \boldsymbol{\Psi}_k), \tag{8.58}$$

where $\mathbf{Z}_k$, $\mathbf{X}_k$, $\mathbf{Q}_k$, $\boldsymbol{\Omega}_k$, and $\boldsymbol{\Psi}_k$ are the variable values at the kth iteration of the ADMM sequence. In a similar manner to the non-RIS-assisted case, the dual variables in $\boldsymbol{\Psi}_{k+1}$, $1 \le m \le M$ are updated via a gradient ascent step, i.e.,

$$\boldsymbol{\Psi}_{k+1} = \boldsymbol{\Psi}_k + \gamma\left(\mathbf{Z}_{k+1} - \mathbf{X}_{k+1}\mathbf{Q}_{k+1}\right). \tag{8.59}$$

We now proceed with the derivation to solution in $(\mathcal{P}_{6A})$, that is,

$$\begin{aligned} \min_{\mathbf{Z}} \; & -\mathrm{tr}(\mathrm{re}(\mathbf{S}^H\mathbf{H}_{ef,k}\mathbf{Z})) + \frac{1}{2}\left\|\mathbf{H}_{ef,k}\mathbf{Z}\right\|_F^2 + \mathrm{tr}(\mathrm{re}(\boldsymbol{\Psi}_k^H\mathbf{Z})) \\ & + \frac{\rho}{2}[\|\mathbf{Z}\|_F^2 - 2\mathrm{tr}(\mathrm{re}(\mathbf{Z}^H\mathbf{X}_k\mathbf{Q}_k))] \\ & \text{s.t. } \frac{1}{M}\mathbf{Z}\mathbf{Z}^H = \mathbf{R}_D, \end{aligned} \tag{8.60}$$

where $\mathbf{H}_{ef,k} = \mathbf{H} + \mathbf{G}_2\boldsymbol{\Omega}_k\mathbf{G}_1$ is the effective channel matrix on the kth iteration. The simplified ,objective function can be written as, $-\mathrm{tr}\{\mathrm{re}(\mathbf{Z}^H(\mathbf{H}^H\mathbf{S} - \boldsymbol{\Lambda}_k + \rho\mathbf{X}_k\mathbf{Q}_k))\}$. This problem is equivalent to $(\mathcal{P}_{2A})$ by replacing $\mathbf{H}$ with $\mathbf{H}_{ef,k}$. Thus, the update of the auxiliary $\mathbf{Z}$ variable at the $(k+1)$th iteration is done by using again (8.43). In addition, problems $\mathcal{P}_{6B}$ and $\mathcal{P}_{2C}$ are identical to $\mathcal{P}_{2B}$ and $\mathcal{P}_{2C}$ and thus, the solutions are again given by (8.45) and (8.49), respectively. It remains to derive the solution to $\mathcal{P}_{2D}$. We have,

$$\min_{\boldsymbol{\Omega}} \frac{1}{2}\|\mathbf{S} - (\mathbf{H} + \mathbf{G}_2\boldsymbol{\Omega}\mathbf{G}_1)\mathbf{Z}_k\|_F^2 \tag{8.61}$$

$$\text{s.t. } \boldsymbol{\Omega} \in \mathcal{F}_D^{N_F\times N_F} \tag{8.62}$$

We once more use an inexact ADMM update to approximate with a closed form the solution to this problem, that is, the unconstrained problem is solved and then the solution is projected onto set ${\mathcal{F}_D}^{N_F \times N_F}$. To that end, it can be shown that the update for $\mathbf{\Omega}$ at the $(k+1)$th iteration is given by

$$\mathbf{\Omega}_{k+1} = \Pi_{\mathcal{F}} \left\{ \text{diag}\{(\mathbf{G}_2^H \mathbf{G}_2)^{-1} \tilde{\mathbf{Z}}_{k+1} (\mathbf{G}_1 \mathbf{Z}_k \mathbf{Z}_k^H \mathbf{G}_1^H)^{-1}\} \right\}, \tag{8.63}$$

where $\tilde{\mathbf{Z}}_{k+1} = \mathbf{G}_2^H(\mathbf{S}\mathbf{Z}_{k+1} + \mathbf{H}\mathbf{Z}_{k+1}\mathbf{Z}_{k+1}^H)\mathbf{G}_1^H$ and the projection operator $\Pi_{\mathcal{F}}\{\cdot\}$ can be calculated as in (8.47).

The termination criteria in (8.53) are again employed to terminate the alternating optimization procedure (8.55)–(8.59). The complete procedure is given in Algorithm 8.2. The computational complexity per iteration is again $O(M^3)$ due to the SVD calculation in the solution of $\mathcal{P}_{6A}$.

Algorithm 8.2 Proposed method for RIS-assisted JRC system's waveform design

Initialize: $\mathbf{X}_0 \in \mathcal{F}^{N_T \times M}$, $\mathbf{Z}_0$ to be matrix satisfying $\frac{1}{M}\mathbf{Z}_0\mathbf{Z}_0^H = \mathbf{R}_D$, $\mathbf{\Omega}_0 \in {\mathcal{F}_D}^{N_F \times N_F}$ and initial $\mathbf{\Psi}_0$ with zeros
while The criteria in (8.52) and (8.53) **do**
  $k \leftarrow k+1$
  Calculate $\hat{\mathbf{S}} = (\mathbf{H} + \mathbf{G}_2\mathbf{\Omega}\mathbf{G}_1)^H\mathbf{S} - \mathbf{\Lambda}_k + \rho\mathbf{X}_k\mathbf{Q}_k$
  Compute the SVD of $\mathbf{F}^H\hat{\mathbf{S}}$ and calculate $\tilde{\mathbf{Z}}$ by (8.42)
  Update auxiliary matrix $\mathbf{Z}_{k+1}$ using solution (8.43)
  Update waveform matrix $\mathbf{X}_{k+1}$ using solution (8.45)
  Update VGA matrix $\mathbf{Q}_{k+1}$ using solution (8.49)
  Update RIS coefficients matrix $\mathbf{\Omega}_{k+1}$ from (8.63)
  Update Lagrange multiplier matrix $\mathbf{\Psi}_{k+1}$ using (8.59)
end while
return $\mathbf{X}_\star, \mathbf{Q}_\star, \mathbf{\Omega}_\star$

8.4.3.1 Simulation Results

We consider a similar setup for generating the simulation results as the non-RIS-assisted wireless system under consideration.

Figures 8.6 and 8.7 examine the performance of the proposed approach for (i) a non-RIS-assisted JRC system, (ii) an 8×8 RIS assisted JRC system, and (iii) a 12×12 RIS assisted JRC system. The other simulation parameters are set to the same values as in the results in Figures 8.3 and 8.4. As observed, both the performance of the communication and the radar parts are significantly enhanced thanks to the deployment of an RIS. For the communication part in Figure 8.6, we observe an increase in the JRC system's sum rate. For the radar

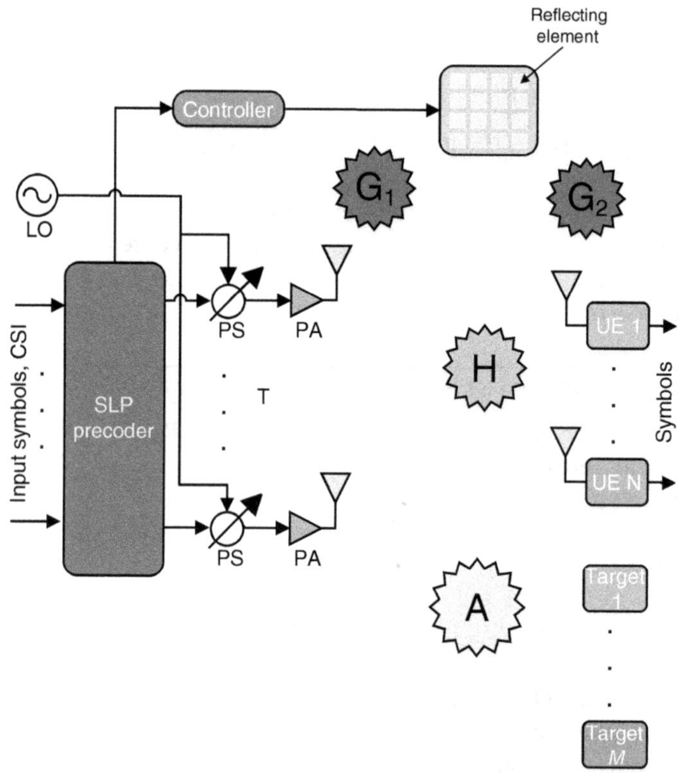

Figure 8.5 An RIS-assisted MIMO JRC system with low complexity analog architecture.

part, observed nulls in the desired beampattern are deeper when an RIS system is employed. As is also expected, when the number of RIS elements increases, the performance is improved for both the communication and the radar parts.

8.5 Conclusion

In this chapter, we study a low-complexity design of waveform for RIS-assisted MIMO JRC system. The chapter starts by introducing the concepts of RIS and JSAC and provides a short survey of the integration of the two potential technologies for next-generation wireless communication system. This is followed by a discussion of the waveform designs studied in the literature for JSAC systems. Following this, we introduced a simplified analog architecture designed to create an efficient RIS-aided MIMO JRC system. To construct a transmit waveform for this JRC system, we formulated a downlink MUI minimization

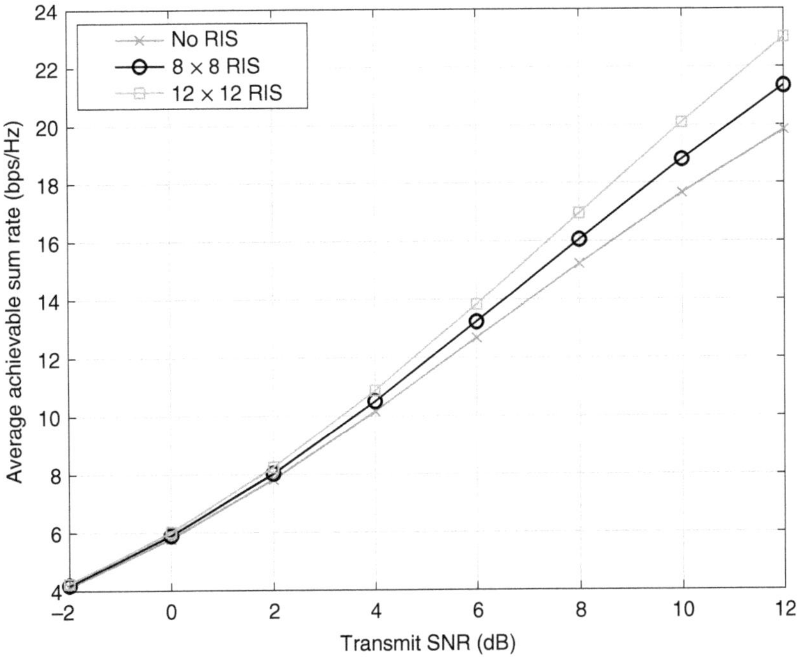

Figure 8.6 Sum rate w.r.t. SNR communication performance comparison for the proposed approach without RIS-assisted and with RIS-assisted communication for an 8×8 and a 12×12 RIS array.

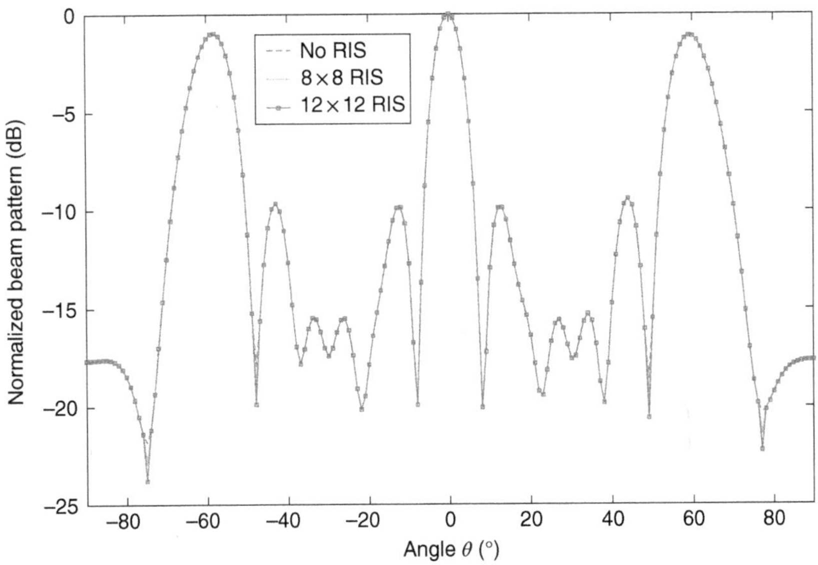

Figure 8.7 Radar beampattern performance for the proposed approach without RIS-assisted and with RIS-assisted communication for an 8×8 and a 12×12 RIS array.

problem, incorporating constraints specific to the architecture and aligned with the desired radar beam pattern. To solve this challenging non-convex optimization problem, we proposed a streamlined solution rooted in the primal-dual framework. The outcomes of our proposed approach are noteworthy, such as achieving substantial gains in spectral efficiency when compared to the current directional-strict baseline. For instance, we observe a remarkable performance boost, approximately a 5 bits/s/Hz rate increase at an SNR of 10 dB. Additionally, our approach demonstrates excellent radar beam pattern performance.

References

1 Ericsson (2022). Ericsson Mobility Report 2022, 1–40. https://www.ericsson.com/4ae28d/assets/local/reports-papers/mobility-report/documents/2022/ericsson-mobility-report-november-2022.pdf (accessed 14 October 2024).

2 3GPP Release 18. https://www.3gpp.org/specifications-technologies/releases/release-18 (accessed 14 October 2024).

3 5G NR Release 17. https://www.qualcomm.com/research/5g/5g-nr (accessed 14 October 2024).

4 Reddy, B.R., Dash, S.P., and Ghose, D. (2023). Optimal multi-level amplitude-shift keying for non-coherent SIMO wireless system in Rician fading environment. *IEEE Transactions on Vehicular Technology* 73 (3): 4493–4498.

5 Reddy, B.R., Dash, S.P., and Ghose, D. (2023). Optimal rotated M-PSK signaling for multi-user receive diversity system over correlated Rician fading channels. *IEEE Communications Letters* 27 (3): 1006–1010.

6 Kundu, N., Dash, S.P., McKay, M.R., and Mallik, R.K. (2022). Channel estimation and secret key rate analysis of MIMO terahertz quantum key distribution. *IEEE Transactions on Communications* 70 (5): 3350–3363.

7 Patil, A., Iyer, S., Lopez, O.L.A. et al. (2023). A comprehensive survey on spectrum sharing techniques for 5G/B5G intelligent wireless networks: opportunities, challenges and future research directions. https://arxiv.org/abs/2211.08956.

8 Dash, S.P., Joshi, S., Shandilya, S.K. et al. (2022). A cybertwin-based 6G cooperative IoE communication network: secrecy outage analysis. *IEEE Transactions on Industrial Informatics* 18 (7): 4922–4932.

9 Basar, E. (2019). Transmission through large intelligent surfaces: a new frontier in wireless communications. *2019 European Conference on Networks and Communications (EuCNC)*, Valencia, Spain, 112–117.

10 Jeganathan, J., Ghrayeb, A., Szczecinski, L., and Ceron, A. (2009). Space shift keying modulation for MIMO channels. *IEEE Transactions on Wireless Communications* 8 (7): 3692–3703.

11 Dash, S.P., Mallik, R.K., and Pandey, N. (2022). Performance analysis of an index modulation-based receive diversity RIS-assisted wireless communication system. *IEEE Communications Letters* 26 (4): 768–772.

12 Pal, R.K., Dash, S.P., Aissa, S., and Ghose, D. (2023). Outage probability analysis of RIS-assisted indoor Wide-FOV-VLC/RF wireless communication. *IEEE Wireless Communications Letters* 13 (1): 34–38.

13 Ajayan, A.P., Dash, S.P., and Ramkumar, B. (2023). Approximate composite channel statistics and performance analysis of IRS-aided wireless system under Nakagami-*m* fading. *IEEE Access* 11: 102290–102300.

14 Singh, R., Kaushik, A., Shin, W. et al. (2024). Indexed multiple access with reconfigurable intelligent surfaces: the reflection tuning potential. *IEEE Communications Magazine* 62 (4): 120–126.

15 Al-Nahhas, B., Nadeem, Q.U.-A., Kaushik, A., and Chaaban, A. (2023). Reconfigurable intelligent surfaces-assisted communication under different CSI assumptions. *2023 IEEE International Mediterranean Conference on Communications and Networking (MeditCom)*, Dubrovnik, Croatia, 18–23.

16 Iqbal, A., Nauman, A., Jamshed, M.A. et al. (2023). Optimizing reconfigurable intelligent surfaces for mmWave communications in IoT networks. *2023 IEEE 34th Annual International Symposium on Personal, Indoor and Mobile Radio Communications (PIMRC)*, Toronto, Canada, 1–6.

17 Dash, S.P., Joshi, S., and Aissa, S. (2022). Envelope distribution of two correlated complex Gaussian random variables and application to the performance evaluation of RIS-assisted communications. *IEEE Communications Letters* 26 (9): 2018–2022.

18 Liu, F., Zhou, L., Masouros, C. et al. (2018). Toward dual-functional radar-communication systems: optimal waveform design. *IEEE Transactions on Signal Processing* 66 (16): 4264–4279.

19 Kaushik, A., Eldar, Y.C., and Dobre, O.A. (2023). Integrated sensing and communications for evolution of next generation networks. *IEEE Communications Technology News (CTN).*

20 Kaushik, A., Singh, R., Dayarathna, S. et al. (2024). Toward integrated sensing and communications for 6G: key enabling technologies, standardization, and challenges. *IEEE Communications Standards Magazine* 8 (2): 52–59.

21 Vlachos, E. and Kaushik, A. (2023). Covariance-based hybrid beamforming for spectrally efficient joint radar-communications. *ICC 2023-IEEE International Conference on Communications*, Rome, Italy, 1–6.

22 Vlachos, E. and Kaushik, A. (2023). Subset selection based RIS-aided beamforming for joint radar-communications. *2023 IEEE Wireless Communications and Networking Conference (WCNC)*, Glasgow, UK, 1–6.

23 Kaushik, A., Singh, R., Li, M. et al. (2024). Integrated sensing and communications for IoT: synergies with key 6G technology enablers. *IEEE Internet of Things Magazine* 7 (5): 136–143.

24 Kaushik, A., Vlachos, E., Thompson, J. et al. (2022). Towards 6G: spectrally efficient joint radar and communication with RF selection, interference and hardware impairments. *IET Signal Processing* 16 (7): 851–863.

25 Tsinos, C.G., Arora, A., Chatzinotas, S., and Ottersten, B. (2021). Joint transmit waveform and receive filter design for dual-function radar-communication systems. *IEEE Journal on Selected Topics in Signal Processing* 15 (6): 1378–1392.

26 Jamshed, M.A., Kaushik, A., Toka, M. et al. (2024). Synergizing airborne non-terrestrial networks and reconfigurable intelligent surfaces-aided 6G IoT. *IEEE Internet of Things Magazine* 7 (2): 46–52.

27 Toka, M., Lee, B., Seong, J. et al. (2024). RIS-empowered LEO satellite networks for 6G: promising usage scenarios and future directions. *IEEE Communications Magazine*. arXiv preprint arXiv:2402.07381.

28 Basu, A., Dash, S.P., Kaushik, A. et al. (2024). Performance analysis of RIS-aided index modulation with greedy detection over Rician fading channels. *IEEE Transactions on Wireless Communications* 23 (8): 8465–8479.

29 Bertsekas, D.P. (1999). *Nonlinear Programming*, 2. Athena Scientific.

30 Vamsidhar, S.R., Dash, S.P., Acharya, R. et al. (2023). Automatic modulation classification in RIS-assisted wireless communication systems using ensemble learning techniques. *2023 IEEE 98th Vehicular Technology Conference (VTC2023-Fall)*, Hong Kong, 1–6.

31 Bhowal, A., Aissa, S., and Dash, S.P. (2023). Irregularly activated spatial modulation schemes with RIS as a modulator. *2023 IEEE 98th Vehicular Technology Conference (VTC2023-Fall)*, Hong Kong, 1–6.

32 Basu, A., Dash, S.P., Joshi, S., and Ghose, D. (2023). Performance of SSK-based receive diversity RIS-assisted system with Nakagami-*m* fading channels. *2023 IEEE 97th Vehicular Technology Conference (VTC2023-Spring)*, Florence, Italy, 1–5.

33 Di Renzo, M., Ntontin, K., Song, J. et al. (2020). Reconfigurable intelligent surfaces vs. relaying: differences, similarities, and performance comparison. *IEEE Open Journal of the Communications Society* 1: 798–807.

34 Basu, A., Dash, S.P., Kaushik, A. et al. (2023). RIS aided index modulation with greedy detection over Rician fading channels. https://doi.org/10.48550/arXiv.2307.09417.

35 Dash, S.P. and Kaushik, A. (2023). RIS-assisted 6G wireless communications: a novel statistical framework in the presence of direct channel. https://doi.org/10.48550/arXiv.2308.06007.

36 Basar, E. (2020). Reconfigurable intelligent surface-based index modulation: a new beyond MIMO paradigm for 6G. *IEEE Transactions on Communications* 68 (5): 3187–3196.

37 Ajayan, A.P., Dash, S.P., and Ramkumar, B. (2022). Performance analysis of an IRS-aided wireless communication system with spatially correlated channels. *IEEE Wireless Communications Letters* 11 (3): 563–567.

38 Li, B., Petropulu, A.P., and Trappe, W. (2016). Optimum co-design for spectrum sharing between matrix completion based MIMO radars and a MIMO communication system. *IEEE Transactions on Signal Processing* 64 (17): 4562–4575.

39 Liu, F., Masouros, C., Li, A. et al. (2018). MU-MIMO communications with MIMO radar: from co-existence to joint transmission. *IEEE Transactions on Wireless Communications* 17 (4): 2755–2770.

40 Zhang, J.A., Liu, F., Masouros, C. et al. (2021). An overview of signal processing techniques for joint communication and radar sensing. *IEEE Journal on Selected Topics in Signal Processing* 15 (6): 1295–1315.

41 Kaushik, A., Vlachos, E., Masouros, C. et al. (2022). Green joint radar-communications: RF selection with low resolution DACs and hybrid precoding. *ICC 2022-IEEE International Conference on Communications*, 1–6.

42 Kaushik, A., Masouros, C., and Liu, F. (2021). Hardware efficient joint radar-communications with hybrid precoding and RF chain optimization. *ICC 2021-IEEE International Conference on Communications*, 1–6.

43 Dizdar, O., Kaushik, A., Clerckx, B., and Masouros, C. (2022). Energy efficient dual-functional radar-communication: rate splitting multiple access, low-resolution DACs, and RF chain selection. *IEEE Open Journal of the Communications Society* 3: 986–1006.

44 Sturm, C. and Wiesbeck, W. (2011). Waveform design and signal processing aspects for fusion of wireless communications and radar sensing. *Proceedings of the IEEE* 99 (7): 1236–1259.

45 Moghaddasi, J. and Wu, K. (2016). Multifunctional transceiver for future radar sensing and radio communicating data-fusion platform. *IEEE Access* 4: 818–838.

46 Kumari, P., Choi, J., González-Prelcic, N., and Heath, R.W. Jr. (2018). IEEE 802.11ad-based radar: an approach to joint vehicular communication-radar system. *IEEE Transactions on Vehicular Technology* 67 (4): 3012–3027.

47 Grossi, E., Lops, M., Venturino, L., and Zappone, A. (2018). Opportunistic radar in IEEE 802.11ad networks. *IEEE Transactions on Signal Processing* 66 (9): 2441–2454.

48 Liu, X., Huang, T., Shlezinger, N. et al. (2020). Joint transmit beamforming for multiuser MIMO communications and MIMO radar. *IEEE Transactions on Signal Processing* 68: 3929–3944.

49 Chen, S., Kaushik, A., and Masouros, C. (2021). Pre-scaling and codebook design for joint radar and communication based on index modulation. *arXiv:2111.10527*, 1–4.

50 Kaushik, A., Thompson, J., Vlachos, E. et al. (2019). Dynamic RF chain selection for energy efficient and low complexity hybrid beamforming in millimeter wave MIMO systems. *IEEE Transactions on Green Communications and Networking* 3 (4): 886–900.

51 Arora, A., Tsinos, C.G., Rao, B.S.M.R. et al. (2020). Hybrid transceivers design for large-scale antenna arrays using majorization-minimization algorithms. *IEEE Transactions on Signal Processing* 68: 701–714.

52 Kaushik, A., Vlachos, E., Tsinos, C. et al. (2021). Joint bit allocation and hybrid beamforming optimization for energy efficient millimeter wave MIMO systems. *IEEE Transactions on Green Communications and Networking* 5 (1): 119–132.

53 Vlachos, E., Thompson, J., Kaushik, A., and Masouros, C. (2020). Radio-frequency chain selection for energy and spectral efficiency maximization in hybrid beamforming under hardware imperfections. *Proceedings of the Royal Society A* 476 (2244): 1–20.

54 Kaushik, A., Vlachos, E., and Thompson, J. (2019). Energy efficiency maximization of millimeter wave hybrid MIMO systems with low resolution DACs. *ICC 2019-2019 IEEE International Conference on Communications (ICC)*, 1–6.

55 Domouchtsidis, S., Tsinos, C.G., Chatzinotas, S., and Ottersten, B. (2019). Symbol-level precoding for low complexity transmitter architectures in large-scale antenna array systems. *IEEE Transactions on Wireless Communications* 18 (2): 852–863.

56 Wang, Z., Liu, Z., Shen, Y. et al. (2022). Location awareness in beyond 5G networks via reconfigurable intelligent surfaces. *IEEE Journal on Selected Areas in Communications* 40 (7): 2011–2025.

57 Shao, X., You, C., Ma, W. et al. (2022). Target sensing with intelligent reflecting surface: architecture and performance. *IEEE Journal on Selected Areas in Communications* 40 (7): 2070–2084.

58 Liu, R., Li, M., Luo, H. et al. (2023). Integrated sensing and communication with reconfigurable intelligent surfaces: opportunities, applications, and future directions. *IEEE Wireless Communications* 30 (1): 50–57.

59 Wang, X., Fei, Z., Huang, J., and Yu, H. (2021). Joint waveform and discrete phase shift design for RIS-assisted integrated sensing and communication system under Cramer-Rao bound constraint. *IEEE Transactions on Vehicular Technology* 71 (1): 1004–1009.

60 Luo, H., Liu, R., Li, M. et al. (2022). Joint beamforming design for RIS-assisted integrated sensing and communication systems. *IEEE Transactions on Vehicular Technology* 71 (12): 13393–13397.

61 Hua, M., Wu, Q., He, C. et al. (2023). Joint active and passive beamforming design for IRS-aided radar communication. *IEEE Transactions on Wireless Communications* 22 (4): 2278–2294.

62 Xing, Z., Wang, R., and Yuan, X. (2023). Joint active and passive beamforming design for reconfigurable intelligent surface enabled integrated sensing and communication. *IEEE Transactions on Communications* 71 (4): 2457–2474.

63 He, Y., Cai, Y., Mao, H., and Yu, G. (2022). RIS-assisted communication radar coexistence: joint beamforming design and analysis. *IEEE Journal on Selected Areas in Communications* 40 (7): 2131–2145.

64 Zhang, H. (2022). Joint waveform and phase shift design for RIS-assisted integrated sensing and communication based on mutual information. *IEEE Communications Letters* 26 (10): 2317–2321.

65 Han, L. and Wu, K. (2013). Joint wireless communication and radar sensing systems-state of the art and future prospects. *IET Microwaves, Antennas and Propagation* 7 (11): 876–885.

66 Kenney, J.B. (2011). Dedicated short-range communications (DSRC) standards in the United States. *Proceedings of the IEEE* 99 (7): 1162–1182.

67 Wymeersch, H., Seco-Granados, G., Destino, G. et al. (2017). 5G mmWave positioning for vehicular networks. *IEEE Wireless Communications* 24 (6): 80–86.

68 Zhang, Q., Sun, H., Gao, X. et al. (2022). Time-division ISAC enabled connected automated vehicles cooperation algorithm design and performance evaluation. *IEEE Journal on Selected Areas in Communications* 40 (8): 2206–2218.

69 Bica, M. and Koivunen, V. (2019). Radar waveform optimization for target parameter estimation in cooperative radar-communications systems. *IEEE Transactions on Aerospace and Electronic Systems* 55 (5): 2314–2326.

70 Shi, C., Wang, F., Sellathurai, M. et al. (2018). Power minimization-based robust OFDM radar waveform design for radar and communication systems in coexistence. *IEEE Transactions on Signal Processing* 66 (5): 1316–1330.

71 Fu, R., Mulleti, S., Huang, T. et al. (2020). Hardware prototype demonstration of a cognitive radar with sparse array antennas. *Electronics Letters* 56: 1210–1212.

72 D'Andrea, C., Buzzi, S., and Lops, M. (2020). Communications and radar coexistence in the massive MIMO regime: uplink analysis. *IEEE Transactions on Wireless Communications* 19 (1): 19–33.

73 Mahal, J.A., Khawar, A., Abdelhadi, A., and Clancy, T.C. (2017). Spectral coexistence of MIMO radar and MIMO cellular system. *IEEE Transactions on Aerospace and Electronic Systems* 53 (2): 655–668.

74 Sodagari, S., Khawar, A., Clancy, T.C., and McGwier, R. (2012). A projection based approach for radar and telecommunication systems coexistence. *2012 IEEE Global Communications Conference (GLOBECOM)*, 5010–5014.

75 Jamil, M., Zepernick, H.-J., and Pettersson, M.I. (2008). On integrated radar and communication systems using Oppermann sequences. *MILCOM 2008-2008 IEEE Military Communications Conference*, 1–6.

76 Ye, Z., Zhou, Z., Fan, P. et al. (2022). Low ambiguity zone: theoretical bounds and Doppler-resilient sequence design in integrated sensing and communication systems. *IEEE Journal on Selected Areas in Communications* 40 (6): 1809–1822.

77 Chen, X., Feng, Z., Wei, Z. et al. (2021). Code-division OFDM joint communication and sensing system for 6G machine-type communication. *IEEE Internet Things Journal* 8 (15): 12093–12105.

78 Mishra, K.V., Eldar, Y.C., Shoshan, E. et al. (2020). A cognitive sub-Nyquist MIMO radar prototype. *IEEE Transactions on Aerospace and Electronic Systems* 56 (2): 937–955.

79 Cohen, D., Mishra, K.V., and Eldar, Y.C. (2018). Spectrum sharing radar: coexistence via Xampling. *IEEE Transactions on Aerospace and Electronic Systems* 54 (3): 1279–1296.

80 Ma, D., Shlezinger, N., Huang, T. et al. (2020). Joint radar-communication strategies for autonomous vehicles: Combining two key automotive technologies. *IEEE Signal Processing Magazine* 37 (4): 85–97.

81 Roberton, M. and Brown, E.R. (2003). Integrated radar and communications based on chirped spread-spectrum techniques. *IEEE MTT-S International Microwave Symposium Digest* 1: 611–614.

82 Saddik, G.N., Singh, R.S., and Brown, E.R. (2007). Ultra-wideband multifunctional communications/radar system. *IEEE Transactions on Microwave Theory and Techniques* 55 (7): 1431–1437.

83 Hassanien, A., Amin, M.G., Zhang, Y.D., and Ahmad, F. (2016). Dual-function radar-communications: information embedding using sidelobe control and waveform diversity. *IEEE Transactions on Signal Processing* 64 (8): 2168–2181.

84 Hassanien, A., Amin, M.G., Zhang, Y.D. et al. (2016). Non-coherent PSK-based dual-function radar-communication systems. *2016 IEEE Radar Conference (RadarConf)*, 1–6.

85 BouDaher, E., Hassanien, A., Aboutanios, E., and Amin, M.G. (2016). Towards a dual-function MIMO radar-communication system. *2016 IEEE Radar Conference (RadarConf)*, 1–6.

86 Huang, T., Shlezinger, N., Xu, X. et al. (2020). MAJoRCom: A dual-function radar communication system using index modulation. *IEEE Transactions on Signal Processing* 68: 3423–3438.

87 Huang, T., Shlezinger, N., Xu, X. et al. (2020). Multi-carrier agile phased array radar. *IEEE Transactions on Signal Processing* 68: 5706–5721.

88 Johnston, J., Venturino, L., Grossi, E. et al. (2022). MIMO OFDM dual-function radar-communication under error rate and beampattern constraints. *IEEE Journal on Selected Areas in Communications* 40 (6): 1951–1964.

89 Keskin, M.F., Wymeersch, H., and Koivunen, V. (2021). MIMO-OFDM joint radar-communications: is ICI friend or foe? *IEEE Journal on Selected Topics in Signal Processing* 15 (6): 1393–1408.

90 Liu, F., Masouros, C., Petropulu, A.P. et al. (2020). Joint radar and communication design: applications, state-of-the-art, and the road ahead. *IEEE Transactions on Communications* 68 (6): 3834–3862.

91 Chen, L., Liu, F., Wang, W., and Masouros, C. (2021). Joint radar-communication transmission: a generalized Pareto optimization framework. *IEEE Transactions on Signal Processing* 69: 2752–2765.

92 Ma, D., Shlezinger, N., Huang, T. et al. (2021). FRaC: FMCW-based joint radar-communications system via index modulation. *IEEE Journal on Selected Topics in Signal Processing* 15 (6): 1348–1364.

93 Zhao, M.-M., Wu, Q., Zhao, M.-J., and Zhang, R. (2021). Intelligent reflecting surface enhanced wireless networks: two-timescale beamforming optimization. *IEEE Transactions on Wireless Communications* 20 (1): 2–17.

94 Wang, J., Wang, H., Han, Y. et al. (2021). Joint transmit beamforming and phase shift design for reconfigurable intelligent surface assisted MIMO systems. *IEEE Transactions on Cognitive Communications and Networking* 7 (2): 354–368.

95 Kaushik, A., Arora, A., Tsinos, C. et al. (2022). Waveform design for joint radar-communications with low complexity analog components. *2022 2nd IEEE International Symposium on Joint Communications & Sensing (JC&S)*, Seefeld, Austria, 1–5.

96 Mohammed, S.K. and Larsson, E.G. (2013). Per-antenna constant envelope precoding for large multi-user MIMO systems. *IEEE Transactions on Communications* 61 (3): 1059–1071.

97 Fuhrmann, D.R. and Antonio, G.S. (2008). Transmit beamforming for MIMO radar systems using signal cross-correlation. *IEEE Transactions on Aerospace and Electronic Systems* 44 (1): 171–186.

98 Zhang, S. and Huang, Y. (2006). Complex quadratic optimization and semidefinite programming. *SIAM Journal on Optimization* 16 (3): 871–890.

99 Tsinos, C.G., Rontogiannis, A.A., and Berberidis, K. (2017). Distributed blind hyperspectral unmixing via joint sparsity and low-rank constrained non-negative matrix factorization. *IEEE Transactions on Computational Imaging* 3 (2): 160–174.

100 Tsinos, C.G., Maleki, S., Chatzinotas, S., and Ottersten, B. (2017). On the energy-efficiency of hybrid analog-digital transceivers for single- and multi-carrier large antenna array systems. *IEEE Journal on Selected Areas in Communications* 35 (9): 1980–1995.

101 Tsinos, C.G. and Ottersten, B. (2018). An efficient algorithm for unit-modulus quadratic programs with application in beamforming for wireless sensor networks. *IEEE Signal Processing Letters* 25 (2): 169–173.

102 Tsinos, C.G., Chatzinotas, S., and Ottersten, B. (2020). Hybrid analog-digital transceiver designs for multi-user MIMO mmWave cognitive radio systems. *IEEE Transactions on Cognitive Communications and Networking* 6 (1): 310–324.

103 Tsinos, C. and Ottersten, B. (2018). An efficient algorithm for unit-modulus quadratic programs with application in beamforming for wireless sensor networks. *IEEE Signal Processing Letters* 25 (2): 169–173.

104 Cevher, V., Becker, S., and Schmidt, M. (2014). Convex optimization for big data: scalable, randomized, and parallel algorithms for big data analytics. *IEEE Signal Processing Magazine* 31 (5): 32–43.

105 Esser, E., Zhang, X.X., and Chan, T.F. (2010). A general framework for a class of first order primal-dual algorithms for convex optimization in imaging science. *SIAM Journal on Imaging Sciences* 3 (4): 1015–1046.

Section IV

9

Blockchain and Smart Contract for Decentralized and Secure Spectrum Management Toward 6G – Beyond Hype

Bikramjit Choudhury[1], Pranav K. Singh[1], Panchanan Nath[1], Ujjal Roy[1], and Anshuman Kalla[2]

[1] *Department of Computer Science & Engineering, Central Institute of Technology Kokrajhar, BTR, Assam, India*
[2] *CGPIT, Uka Tarsadia University (UTU), Bardoli, Gujarat, India*

9.1 Introduction

Sixth generation (6G) is emerging on the horizon, poised to advance upon the groundwork established by fifth generation (5G). As shown in Figure 9.1, 6G promises an array of enhancements over fourth generation (4G) and 5G, including greater reliability, ultramassive connectivity, ultrahigh precision, remarkably low energy consumption, cost-efficiency, and reduced latency [1, 2]. Additionally, 6G will introduce heightened intelligence and speeds that are nearly 100 times faster than its predecessor [3]. These advancements have the potential to transform economies and lives globally, all while prioritizing sustainability and eco-friendly initiatives [4].

In order to accommodate a wide range of Quality of Service (QoS) and Quality of Experience (QoE) requirements, 6G is poised to integrate a combination of evolutionary technologies throughout its architectural framework as shown in Figure 9.2. 6G will leverage existing frequencies, ranging from low to high bands, while also exploring new spectrum bands like the THz band and visible light bands (VL) as depicted in Figure 9.3 [9]. In order to utilize the true potential of 6G technology, it is anticipated that ultrahigh data rates reaching terabits per second (Tbps) will be necessary. Consequently, the exploration and utilization of novel frequency bands in the THz and visible light spectrum, as well as the facilitation of dynamic spectrum sharing (DSS), become indispensable strategies for advancing 6G [7, 8, 10].

Intelligent Spectrum Management: Towards 6G, First Edition.
Edited by Sridhar Iyer, Anshuman Kalla, Onel Alcaraz López, and Chamitha De Alwis.

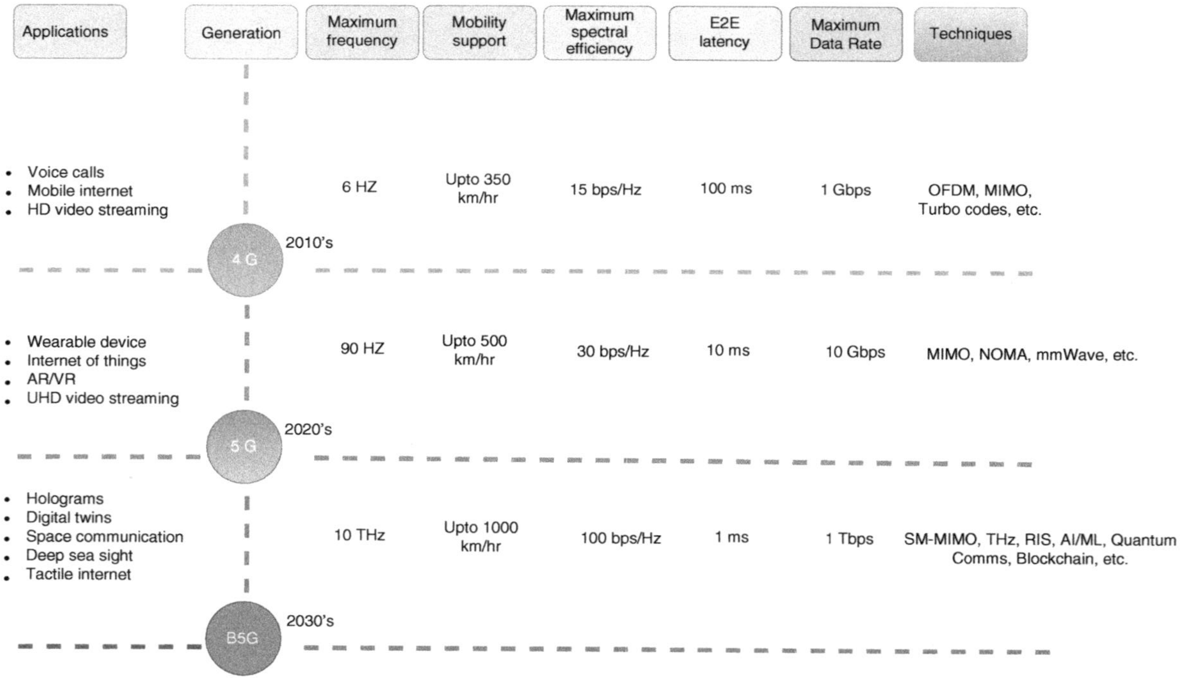

Figure 9.1 The evolution of mobile networks from 4G to beyond 5G (B5G). Source: Adapted from [5].

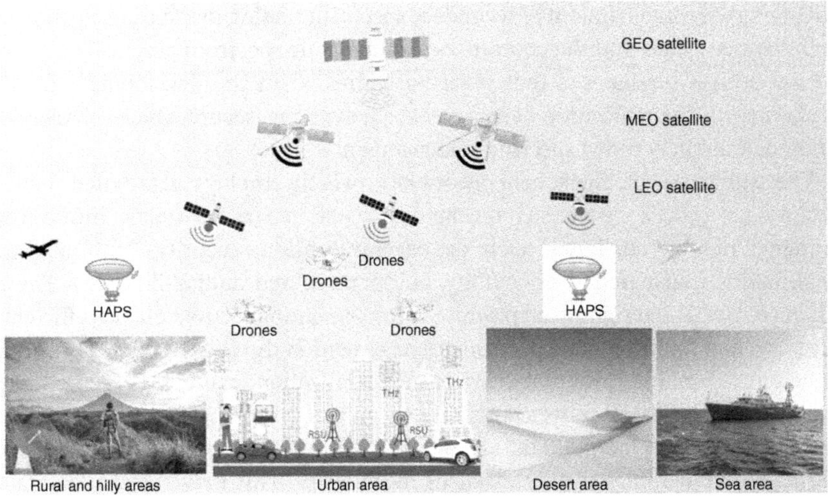

Figure 9.2 6G architecture. Source: Adapted from [11].

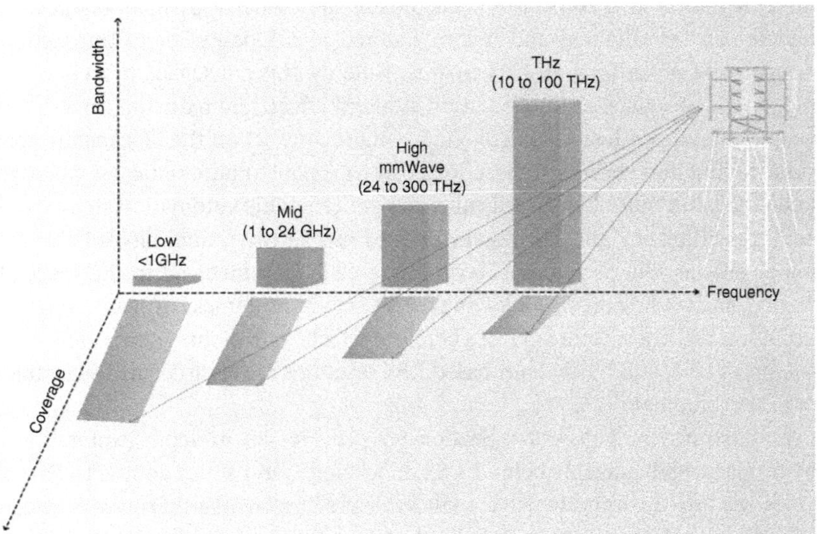

Figure 9.3 6G spectrum. Source: Adapted from [4, 11].

The implementation of dynamic spectrum management primarily relies on spectrum-sharing schemes that harness spectrum databases (use of geolocation databases) and cognitive radio techniques. It introduces concerns related to reliability, security, and privacy in the context of spectrum sharing. Additionally, there is a pressing need to support other aspects of dynamic spectrum management, including the establishment of secondary spectrum trading markets, mitigation of interference issues and the coordination of dynamic spectrum access. Thus, one of the critical bottlenecks that must be addressed in the development of 6G technology is the realization of decentralized, dynamic, secure, efficient, reliable, trusted, and finely tuned spectrum management.

The utilization of blockchain technology, which employs distributed ledger technology (DLT), consensus mechanisms, and smart contracts, introduces a range of key attributes such as decentralization, security, immutability, traceability, transparency, anonymity, automation, and auditability [12]. These characteristics hold significant promise in implementing secure, reliable, efficient, scalable, and fault-tolerant spectrum management in the context of 6G [13]. DLT inherently combines immutability and decentralization, making it exceptionally suitable for recording transactions among parties who do not inherently trust each other. It has the potential to establish a distributed peer-to-peer network through cryptography and a consensus mechanism. This network can facilitate secure spectrum sensing, spectrum resource auctioning, trading, and spectrum access and regulation without the need for a trusted third party [14]. Through the implementation of a consensus mechanism, the ownership rights of spectrum holders can be validated and securely stored in a block prior to transactions, ensuring the revenue of all participants. Notably, this consensus process can be integrated with a spectrum allocation strategy, effectively reducing unnecessary computational overhead. By deploying a smart contract on the blockchain, spectrum trading and sharing transactions can be executed automatically, following predefined and mutually agreed rules and policies. This automation significantly enhances efficiency and fairness in the process. Furthermore, blockchain technology enables the permanent recording of each step involved in the spectrum sharing, and trading process allowing for the tracing and tracking of any violations through transaction records [15]. Therefore, blockchain technology is well suited to support DSS, and blockchain-based DSS solutions are regarded as key enablers for 6G development [16, 17].

As illustrated in Figure 9.4, Section 9.2 provides an in-depth exploration of the fundamental concepts behind DSS, blockchain, and smart contracts. Section 9.3 delves into a comprehensive analysis of previous works that have leveraged blockchain and smart contract technologies in the context of spectrum sharing for 5G networks, along with a discussion of their limitations. In Section 9.4, we introduce a framework centered around blockchain and smart contracts for spectrum sharing in the context of B5G and 6G networks. The ensuing section, Section 9.5, presents the pertinent challenges associated. Finally, our chapter culminates in Section 9.6, which offers the concluding insights and remarks.

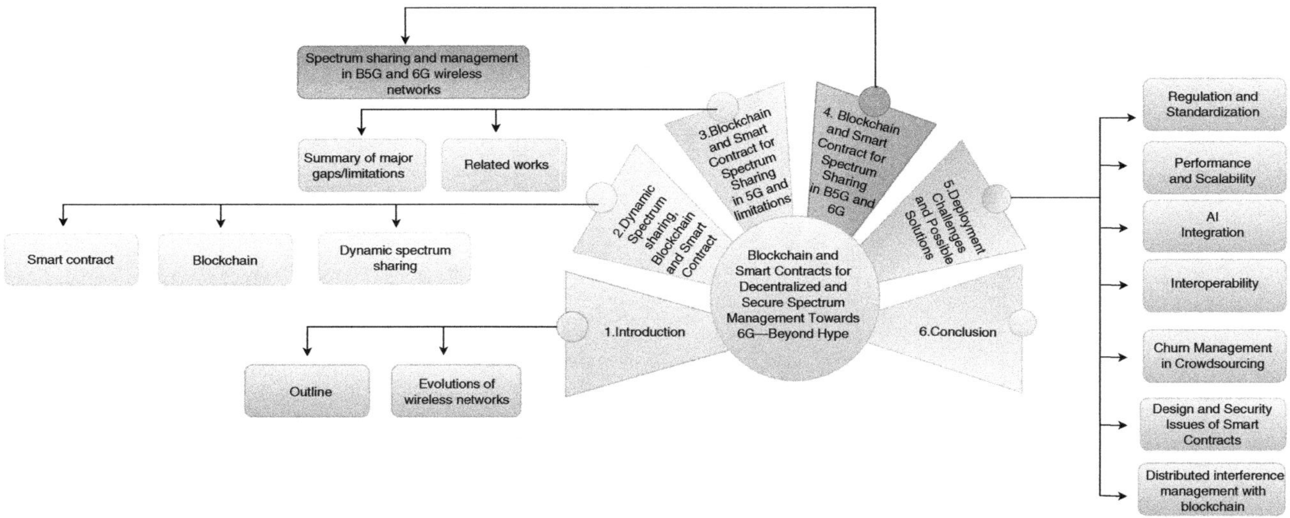

Figure 9.4 Outline of the sections.

9.2 Dynamic Spectrum Sharing, Blockchain, and Smart Contract

In this section of the chapter, we provide details of DSS, blockchain technology, and smart contracts.

9.2.1 Dynamic Spectrum Sharing

DSS facilitates the transfer of allocated spectrum with the goal of enhancing the utilization of the scarce spectrum and meeting the growing demands of the users. There are mainly two types of DSS:

- **Centralized DSS:** In the centralized approach, advanced management nodes (AMNs) are introduced to reduce interference among operators [6], managing and allocating shared spectrum resources in a unified manner. As depicted in Figure 9.5a, each operator comprises one operation administration and maintenance (OAM) server and multiple base stations (BSs). The OAM server is responsible for spectrum assignment and BS management. During DSS, each BS initially sends a spectrum demand request to its respective OAM server. The OAM server evaluates the spectrum demand and submits these demands to the AMN for obtaining spectrum resources. The AMN then takes allocation decisions and communicates the results back to the OAM servers. Subsequently, each BS can dynamically adjust its operating frequency under the OAM server's control. In the centralized DSS architecture, spectrum resource management can occur at either the operator level or the BS level.
- **Decentralized DSS:** Instead of a common AMN node, in the decentralized architecture, each mobile network operator (MNO) has its spectrum controller, either as an independent physical entity or a functional module. These controllers, integrated into entities like OAM servers, interact through predefined communication interfaces as depicted in the Figure 9.5b. Each BS reports spectrum requests to its spectrum controller, including traffic load and service requirements. Based on this data spectrum controller calculates spectrum demand and communicates with other spectrum controllers for available channels and power. Available channels are then allocated to BSs, which configure them for associated users.

Although both centralized and decentralized DSS approaches exist, they face limited adoption due to the absence of proper incentive mechanisms, QoS assurances, and concerns about security and privacy. To enable flexible and efficient spectrum management, emerging technologies like AI and blockchain are essential in the context of 6G networks. Some of the key challenges in the

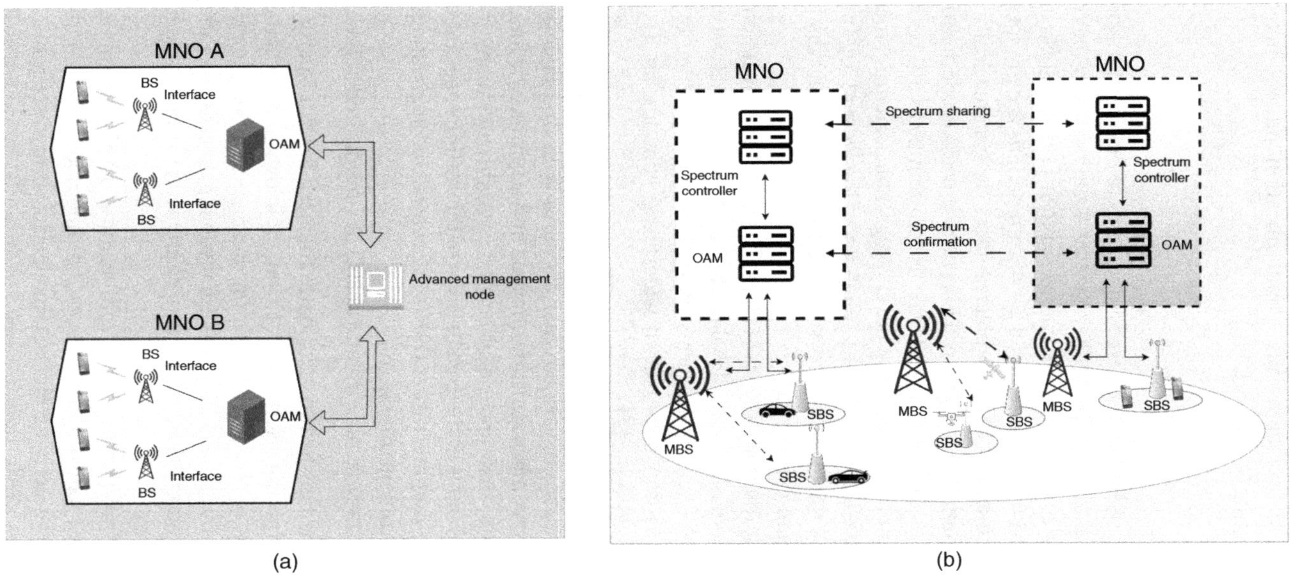

Figure 9.5 DSS architecture. Source: [6]/Springer Nature: (a) centralized and (b) decentralized.

effective implementation of DSS are sensing capabilities, reliability of sensing, interoperability, regulatory approvals, communication among heterogeneous stakeholders, and testing framework [18].

9.2.2 Blockchain

Blockchain is a type of DLT that consists of a series of blocks connected to each other using a hash chain. It essentially serves as a digital record or transaction repository accessible to all computers connected to it [19]. In blockchain, every event is recorded as a transaction, and a set of transactions are grouped together into a block. Subsequently, each completed block is systematically linked to others to form a sequential blockchain. This entire process is commonly referred to as mining. Each record within the ledger is timestamped with both the time and date and is accompanied by a unique cryptographic signature, meticulously designed to ensure the authenticity and reliability of the ledger [20]. Importantly, all participants enjoy unrestricted access to the complete ledger, which provides a comprehensive historical record.

There are primarily four major types of blockchains: public, private, consortium, and hybrid; classified based on the level of accessibility to the network. Over the years, a variety of blockchain platforms have been developed and examples of the leading platforms that have gained widespread adoption are Hyperledger, Ethereum, Corda, and Ripple [21].

While blockchain first gained prominence with the cryptocurrency Bitcoin, it quickly found applications in various other domains. With its unique and desired characteristics blockchain has radically changed and improved the functioning of banking, supply chain, insurance, healthcare, copyright, and many more areas. Furthermore, blockchain is at the core of some of the cutting edge and innovative applications like non-fungible tokens (NFTs), decentralized applications (dApps), virtual marketplace, and metaverse.

Blockchain possesses several key features [16] as summarized below:

- **Decentralization:** This feature eliminates the need for third-party validation of transactions, granting users control over their own data. Furthermore, it holds the potential to obviate the need for database administrators. Spectrum Sharing is inherently distributed in nature involving diverse stakeholders in an untrusted environment thus making it perfectly suitable for blockchain-based solution.
- **Transparency:** Transparency is a core feature of blockchain technology. With its distributed ledger spread across multiple nodes, the complete transaction history and underlying blockchain algorithms are accessible for review by any component. This transparency significantly improves visibility and contributes to the efficient enforcement of spectrum management rules.

- **Immutability:** Immutability is a fundamental aspect of blockchain technology, characterized by its inherent resistance to alteration once data is recorded. This immutability attribute is of great importance in ensuring precise auditing, enforcement, and compliance with regulatory or contractual requirements.
- **Availability:** The sharing of Blockchain ledgers among nodes significantly enhances the availability of information stored on the blockchains. This elevated level of availability transforms blockchain into a dependable spectrum management database, effectively safeguarding incumbent users.
- **Security:** Security is a paramount feature of blockchain technology. Entries in the blockchain ledger are safeguarded through cryptographic measures, and all node-to-node transactions are digitally signed. Users utilize private/public cryptographic key pairs to access the blockchain, thus fortifying the security of wireless network infrastructure against potential threats and attacks.

By embracing these above fundamental features, blockchain emerges as a compelling technology for the future of 6G mobile networks, promising enhanced security, transparency, and efficiency. As a result, in this chapter, we delve into the pivotal role of blockchain as a foundational technology for enabling automated, decentralized spectrum trading and mutual agreements across different entities within a mobile network [15].

9.2.3 Smart Contract

A smart contract is a program coded that operates on a blockchain network environment to facilitate, execute, and automatically enforce the terms of an agreement when specific conditions are met [22, 23]. The concept of a smart contract was initially introduced in the 1990s as a digital transaction protocol designed to autonomously execute the terms of a contractual agreement [24]. Since 2008, when blockchain technology came into existence through Bitcoin cryptocurrency, the importance of smart contract integration has become a focus area to support the diversified applications of blockchain.

Smart contracts are essentially implemented on top of blockchain networks, where approved contractual clauses are translated into executable computer programs. The logical relationships between these clauses are preserved in the form of programmatic flows, often using constructs like if-else-if statements. Each action or statement within a contract's execution is recorded as an immutable transaction, permanently stored on the blockchain ledger.

Smart contracts provide robust access control and contract enforcement mechanisms. Developers have the capability to designate access permissions for each function within a contract. When any condition outlined in a smart contract is satisfied, the corresponding function is automatically executed in a predictable and immutable manner.

9.3 Blockchain and Smart Contract for Spectrum Sharing in 5G

This section of the chapter provides an overview of related work in the field of blockchain and smart contract-enabled spectrum sharing for 5G and beyond. It also highlights the limitations of existing research and identifies research gaps in this area.

9.3.1 Related Works

In a study conducted by Maksymyuk et al. [25], the authors advocated for the synergistic integration of blockchain and AI in the intelligent management of 5G networks. Authors introduced a novel open business model designed to facilitate two key functionalities: (i) the dynamic leasing of spectrum resources by MNOs from government entities that govern spectrum allocation and (ii) the creation of a streamlined pathway for new entrants to join the MNO landscape by leasing unlicensed spectrum resources from established network operators. To ensure equitable and efficient spectrum access while maintaining fair conduct among network operators, the study proposed the concept of spectrum tokenization. The authors conducted comprehensive numerical simulations in a scenario featuring two operators, leveraging the 5 GHz industrial, scientific and medical (ISM) band, to showcase the effectiveness of their proposed algorithm in achieving a harmonious balance between data throughput and the distribution of spectrum tokens over time. It is noteworthy, however, that the study did not delve into the practical implementation of a blockchain-based framework in a multioperator environment, leaving this avenue for further exploration and investigation.

In [16], Weiss et al. explored the utility of blockchain technology in radio spectrum management, with a particular focus on its role in the realm of DSS. The authors underscored several fundamental attributes of blockchain technology, including decentralization, transparency, immutability, availability, and security, and elaborated on how these features could be harnessed to enhance various facets of spectrum management. Decentralization was identified as a pivotal factor that could obviate the necessity for intermediaries like spectrum licensees and band managers. The transparency inherent to blockchain was shown to provide real-time insight into the dynamics of spectrum sharing (SS). The immutability aspect of blockchain was emphasized for its capacity to ensure the integrity of records, making them resistant to tampering. Furthermore, the heightened security conferred by blockchain technology was discussed in the context of reducing fraudulent activities within the SS process. The study also placed significant emphasis on the deployment of smart contracts as a means of transparently and trustlessly enforcing intricate SS rules. Within the study, SS was categorized into primary cooperative, primary noncooperative, secondary cooperative, and secondary noncooperative sharing, and an extensive analysis was provided regarding the advantages and drawbacks of implementing blockchain technology within each of these categories. However, this study has not demonstrated the blockchain technology integration for radio spectrum management.

Li et al. [14] introduced a consortium blockchain-based approach designed to establish an efficient and secure framework for DSS in 5G networks. This approach empowered regulatory bodies with the capacity to oversee the entirety of the DSS process, thereby ensuring equitable revenue distribution among all participants involved. MNOs operating within the blockchain network were able to dynamically switch roles, serving either as spectrum providers or requestors based on the prevailing demand. To facilitate transparent and immutable record-keeping of spectrum allocation, the study leveraged smart contracts, which were recorded on the blockchain ledger. The authors further delved into the intricacies of optimal spectrum pricing and purchase strategies, employing a multileader multifollower Stackelberg game model to address these critical aspects. In practical implementation, the authors constructed a working prototype utilizing the hyperledger fabric framework. This prototype was subsequently used to assess the mean delay, demonstrating the feasibility and scalability of the blockchain-based DSS method in real-world scenarios. This work introduced a DSS system in which multiple operators collaborate through a consortium blockchain to achieve secure and efficient sharing of spectrum resources between them. It is important to note that this study does not address the heterogeneous configuration involving both public and private blockchains for DSS, which is a crucial aspect to consider for its practical implementation in real-world scenarios.

In [26], the authors investigated the synergistic potential of blockchain technology and software-defined networking (SDN) for the purpose of dynamic spectrum management. Their study unveiled a novel framework empowered by blockchain, designed to manage spectrum resources within an integrated network. This comprehensive framework encompassed both interslice SS and intraslice spectrum allocation. The interslice SS component relied on a consortium blockchain, established and maintained by upper-tier SDN controllers. In contrast, the intraslice spectrum assignment was achieved through the utilization of a graph coloring-based channel assignment algorithm. To ensure the integrity and efficiency of the consortium blockchain, the study introduced a bilateral confirmation protocol and a consensus mechanism. Notably, the consensus mechanism exhibited a noteworthy advantage in terms of speed when compared to practical Byzantine fault tolerance algorithms, resulting in a faster consensus-reaching capability. Furthermore, the authors demonstrated that the proposed channel allocation method yielded significant improvements in spectrum utilization in comparison to existing methods. This highlighted the promise of their blockchain-enabled framework in enhancing the effective utilization of available spectrum resources. But, they did not conduct an evaluation of their scheme using a real blockchain platform. Additionally, the study did not take into account the critical aspect of inter-MNO SS, which is essential for realizing the practical implementation and effectiveness of their proposed framework.

In [27], Wang and Sun provided an overview of the standardization progress concerning SS and the utilization of blockchain technology in wireless networks. They also highlighted the critical challenges faced by techniques that enable SS through blockchain and put forward potential solutions. The paper concluded by delineating avenues for future research in the realm of blockchain-enabled DSS. This study is a concise survey that provides an overview of the standardization initiatives related to SS and blockchain technology across multiple standards developing organizations (SDOs). However, it's important to note that this study does not delve into the specifics of blockchain integration or provide practical demonstrations of its implementation.

Gorla et al. [28] introduced a blockchain-based model for multioperator service provisioning to 5G users. Their primary focus was on the management of both intra and interspectrum resources among telecom operators. They presented a blockchain-enabled approach to SS among operators, aiming to reduce underutilization of available spectrum. The research also evaluated the time-sensitive operations of the proposed blockchain framework, emphasizing its capacity to offer transparency, security, and efficient SS operations concurrently. Nonetheless, the study does not address the issues of scalability and fairness within the SS process, particularly in the context of 5G applications. Summary of the aforementioned works is listed in Table 9.1.

9.3.2 Summary of Major Gaps/Limitations

- **Limited Multioperator Blockchain Deployment:** One significant limitation observed in the reviewed literature is the absence of practical blockchain deployment in multioperator environments. This highlights the imperative need to address real-world implementation challenges in such complex scenarios.
- **Incentive Mechanism:** To motivate users or operators with underutilized spectrum resources to participate in DSS, it is crucial to establish effective incentive systems [29]. In traditional DSS setups, these incentives are typically managed by a central authority. However, this approach may lead to imbalances between parties, potentially affecting their revenue. Leveraging blockchain technology, we can explicitly define obligations and responsibilities within smart contracts, thereby ensuring that the expected rewards are received.

Table 9.1 Summary of related works.

References	Year	Objectives	Methodology	Limitations
Maksymyuk et al. [25]	2019	Integration of blockchain and AI in 5G management	Spectrum tokenization	Lack of practical blockchain implementation
Weiss et al. [16]	2019	Utility of blockchain in spectrum management	Categorization of SS approaches	No practical blockchain integration demonstrated
Li et al. [14]	2022	Consortium blockchain for DSS in 5G networks	Stackelberg game model	Heterogeneous blockchain configuration not addressed
Li et al. [26]	2022	Synergy of blockchain and SDN for spectrum management	Inter-slice and intra-slice allocation	Lack of evaluation on a real blockchain platform; Neglects inter-MNO spectrum sharing
Wang and Sun [27]	2022	Overview of standardization progress in spectrum sharing and blockchain	Blockchain-enabled dynamic Spectrum Sharing	Lack of details on blockchain integration and practical demonstrations
Gorla et al. [28]	2020	Blockchain-based model for multioperator 5G service provisioning	Blockchain-enabled spectrum sharing	Scalability and fairness issues not addressed

- **Inadequate Emphasis on Scalability and Fairness:** While the studies briefly touch upon blockchain scalability in SS solutions, they do not delve deeply into the potential challenges related to scalability and fairness, especially in the context of deploying blockchain in large-scale 5G networks.
- **Insufficient Exploration of Real-World Implementations:** Despite referencing several studies, the literature reviewed does not provide comprehensive insights into real-world implementation outcomes or case studies. This gap in understanding practical applications and the adoption of blockchain in these contexts leaves room for further investigation.
- **Lack of Comprehensive Security Evaluation:** Recognizing the security benefits of blockchain, the reviewed studies fall short in offering a thorough assessment of the security mechanisms employed in blockchain-based solutions. A more detailed examination of how blockchain addresses security concerns, especially in DSS scenarios, will provide valuable insights.

- **Need for Standardization:** Although the reviewed literature briefly mentions standardization efforts related to SS and blockchain in wireless networks, it does not delve into the challenges and the critical role of industry-wide standards in ensuring interoperability and consistency in blockchain-based solutions.
- **Minimal Discussion of Regulatory Challenges:** The studies do not address potential regulatory hurdles and legal considerations that may arise during the implementation of blockchain-based solutions in sectors such as 5G networks and SS. Establishing legal and regulatory frameworks is crucial for the successful deployment of these technologies.

9.4 Blockchain and Smart Contract for Spectrum Management in B5G and 6G

In this section, we explore the integration of blockchain and smart contracts in the context of SS, specifically within B5G and 6G networks.

9.4.1 Spectrum Management from B5G and 6G Perspective

In the context of B5G and 6G networks, there is a growing demand for efficient spectrum allocation, sharing, and management. At the heart of this is the notion of decentralized and dynamic spectrum management, which aims to optimize spectrum utilization without compromising network performance and security. This approach revolves around the dynamic and automated allocation of available spectrum resources, enabling multiple users, including both established operators and new entrants, to access the same frequency bands while coexisting seamlessly.

The fundamental idea is to enhance spectral efficiency, reduce interference, and boost overall network capacity through the utilization of advanced technologies like blockchain and AI [30]. As we transition into the era of 6G, these principles remain pertinent, with a heightened focus on ultralow latency and massive connectivity. In the context of 6G networks, secure, efficient, private, trusted, transparent, scalable, fair, and interference-free SS and management holds paramount significance [31]. Blockchain and smart contract technologies offer the means to fulfill these prerequisites, thereby laying the groundwork for the widespread adoption of emerging applications including autonomous vehicles, immersive extended reality, virtual reality, digital twins, metaverse, tactile Internet, and many more.

Blockchain and smart contract have the potential to play a pivotal role in several facets of spectrum management, spanning spectrum allocation, sensing, sharing, spectrum access coordination, service level agreement, and policy enforcement. Additionally, they offer adaptable solutions for other integral components of dynamic spectrum management, including the establishment of secure, immutable geolocation databases, and the facilitation of secondary spectrum trading markets. In this section of the chapter, we explore the diverse applications of blockchain and smart contracts within the realm of spectrum management, specifically in the context of B5G and 6G networks. We also delve into the relevant research and related works within this field.

9.4.2 Blockchain and Smart Contract for Spectrum Allocation

Up to the 5G era, spectrum allocation to telecom providers followed a fixed allocation model, with a central authority or government granting licensed spectrum to organizations. However, this fixed spectrum allocation system suffered from several limitations, including challenges in ensuring a fair allocation process, delays in auction procedures, and the potential for collusive bidding, largely due to its centralized nature.

In the context of 6G-based spectrum band allocation, the focus needs to shift toward guaranteeing QoS and high availability for a diverse range of user applications. The entire spectrum allocation process must be characterized by fairness, security, efficiency, and trustworthiness.

To tackle these spectrum allocation challenges, blockchain and smart contracts emerge as promising solutions. Blockchain technology can effectively address the inherent issues of unfair or biased allocations, and it accomplishes this by eliminating the need for intermediaries through the utilization of smart contracts. With blockchain, it becomes possible to establish a transparent, efficient, secure, and fair system for allocating spectrum to mobile operators. This dynamic and cost-effective approach enhances the overall allocation process, ensuring that it aligns with the evolving demands of the 6G era.

9.4.3 Blockchain and Smart Contract for Spectrum Sensing

Spectrum sensing is crucial for enabling the efficient and flexible use of the radio spectrum, mitigating interference, improving network capacity, and promoting coexistence between different wireless systems. Various techniques, including energy detection, matched filtering, cyclostationary feature detection, compression, weighting, deep sensing, statistical analysis, and ML-based techniques, are used for spectrum sensing [32]. Blockchain and smart contracts can be leveraged to implement secure, reliable, trusted, distributed, and cooperative spectrum sensing across different bands in 6G networks. For instance, blockchain and smart contract-enabled cooperative spectrum sensing can be designed as follows [33]:

- Each spectrum sensor senses for usable vacant frequencies and then broadcasts its sensing results to other nodes within the blockchain network.

- These sensing results are subsequently incorporated as a transaction block within the blockchain via a mining process.
- When a user desires to access an available channel, they trigger a smart contract that enables them to place a bid for that particular channel.
- A module within the smart contract is responsible for selecting the winning bid for the available channel, which then triggers the transfer of payments to miners and the transfer of spectral usage rights to the bid's winner.

9.4.4 Blockchain and Smart Contract for Spectrum Sharing and Trading

In the context of DSS, there are four distinct schemes for achieving SS, which can be categorized as follows: interoperator, intraoperator, crowdsourcing sensing, and trading-based. While these schemes perform admirably concerning various use cases, they all face different challenges concerning security, privacy, and transparency throughout their execution. To address these concerns and enhance the transparency, trustworthiness, and security of the spectrum-sharing process, blockchain technology can be leveraged. Furthermore, the entire process can be automated through the implementation of smart contracts [10].

The centralized spectrum database-based approach for spectrum trading relies on administrative examination and approval mechanisms. However, this approach faces several challenges, including security, trust issues, and high administrative and maintenance costs. In response to these challenges, various economic models have been proposed within centralized database-assisted systems, such as the auction model, spot market model, contract model, and commodity model. Nevertheless, these models also encounter issues related to currency and payment platforms, the establishment of universal standards, bidding techniques, economic models, support for QoS, and policy enforcement, among other concerns [34].

The inherent characteristics and features of blockchain and smart contracts can be harnessed to address the challenges associated with spectrum trading. Blockchain has demonstrated its capacity in the realm of cryptocurrency-based financial systems, making it a viable solution for establishing a robust spectrum payment system using digital currency that can be readily converted into money. The functions of a geo-location database and the requirements of spectrum administrators can be seamlessly integrated into a blockchain platform. Additionally, the bidding language, economic model, primary user protection requirements, QoS needs of users, and other contract enforcement modules can all be effectively deployed on blockchain platforms through the use of smart contracts.

Blockchain technology can play a pivotal role in providing a secure and trustworthy platform for both buyers and sellers within the spectrum trading system.

Sellers can leverage smart contracts to advertise their available channels on the blockchain-based platform. Buyers interested in accessing these available channels can then submit bids within designated time slots, utilizing their cryptocurrency wallets to place bids for spectrum access. The allocation of spectrum leases and usage rights to the winning bidders can be achieved through a consensus mechanism and smart contracts, ensuring a transparent and efficient process.

9.4.5 Blockchain and Smart Contract for Spectrum Access Coordination

Spectrum access coordination (SAC) plays a pivotal role in the management and optimization of spectrum utilization in the context of 6G, with the aim of enhancing spectrum efficiency, QoS, and mitigating interference and collisions. This necessitates the establishment of a set of guidelines or policies that mandate collaborative and coordinated efforts for spectrum access. SAC can be realized by facilitating the exchange of spectrum access requests and achieving a consensus on spectrum access through the use of smart contract and DLT.

9.4.6 Blockchain and Smart Contract for Spectrum Regulation

To harness the full advantages of DSS in 6G, it is imperative to implement robust and thorough spectrum regulation. This entails evaluating spectrum utilization, effectively managing interference and jamming, enabling privacy and maintaining stringent security operations. Spectrum regulation must ensure the unhindered operation of authorized communications while preventing illegal and unintended signal transmissions. Yet, in the context of a heterogeneous wireless network with an SAC integrated architecture in 6G, spectrum regulation becomes an exceptionally complex and challenging endeavor. By harnessing the power of blockchain technology and smart contracts, it becomes possible to monitor and record comprehensive spectrum utilization. The process also includes securely storing spectrum sensing results in an immutable format, logging and tracing signal and interference identification, and facilitating the resolution of disputes. Through this approach, a fully traceable, end-to-end supervision mechanism can be established, capitalizing on the tamper-proof characteristics of the blockchain.

9.4.7 Blockchain and Smart Contract for Service-Level Agreements

A blockchain and smart contract-based framework can effectively bring together MNOs, mobile virtual network operators (MVNOs), and regulatory bodies to enhance dynamic spectrum management in 6G. Smart Contracts can be utilized to implement service level agreements (SLAs) on the blockchain, following consensus on regulatory policies established by the regulatory body and the SS conditions between MNOs and MVNOs. This ensures that subscribers of participating MVNOs receive a higher QoS compared to those on opportunistic MVNOs. Such agreements also facilitate seamless handovers and roaming for subscribers within MNOs and between different MNOs. These SLAs enable dynamic switching between MNOs when an operator's service is unavailable. Smart contracts on this unified platform not only implement the SS agreements but also manage the complexities of SLAs among all stakeholders. Furthermore, these smart contracts can automate the billing process.

9.5 Deployment Challenges and Possible Solutions

Although blockchain holds promise for the advancement of 6G technology, it is essential to recognize and address the challenges that may hinder its full potential in this domain. This section delves into a comprehensive examination of the deployment challenges and offers potential solutions to overcome those challenges.

9.5.1 Regulation and Standardization

SS, in the context of 6G networks, presents a series of regulatory and standardization challenges that must be addressed for effective and efficient operation. As 6G is still evolving, standardization is still a work in progress and protocols and regulations are being developed. Third Generation Partnership Project (3GPPP) is the nodal body for standardization while International Telecommunication Union (ITU) is finalizing spectrum regulations. The challenges encompass the dynamic nature of 6G networks, the management of interference, diversified applications and devices, and the security of critical applications. Standardization plays a pivotal role in mitigating these challenges by providing adaptable regulatory frameworks [10]. Furthermore, the integration of crowdsourced data, guided by standardized collection and processing methods, can enhance the accuracy and reliability of SS.

As blockchain technology gains prominence in this space, standardization efforts should focus on addressing issues such as scalability, interoperability, and privacy protection, ensuring transparent, and secure integration. Collaborative efforts among industry stakeholders, regulatory bodies, and standards organizations are pivotal in addressing these challenges and establishing the necessary standards and protocols for an effective SS ecosystem in the era of 6G.

9.5.2 Performance and Scalability

One key challenge is achieving high-performance spectrum allocation and management in real time, especially in dynamic 6G environments where resources must be allocated swiftly to meet varying demands. The trade-off here may involve the complexity of allocation algorithms versus the time required for decision-making. One possible solution is the development of advanced ML and optimization algorithms that can strike a balance between resource allocation efficiency and real-time responsiveness.

Scalability is another challenge, particularly in accommodating the growing number of devices and services in 6G networks. Ensuring that the SS framework can scale to meet these demands without sacrificing performance is crucial. A solution lies in the implementation of distributed and edge computing architectures that can handle resource allocation at scale while minimizing latency [18]. Moreover, the use of blockchain technology, despite its potential benefits, can introduce scalability challenges in terms of transaction processing speed. Solutions may include leveraging off-chain techniques for faster transaction processing while using the blockchain for security and transparency.

9.5.3 AI Integration

The integration of AI into SS for 6G networks, particularly in the domains of reputation management and incentivization, presents both challenges and potential solutions. Challenges include the need for AI algorithms to accurately assess the reputation of network entities, such as MNOs, amidst the dynamic nature of 6G networks. To address this, AI-driven models should continually adapt and learn from real-time data, enhancing the accuracy of reputation evaluations. Incentivization mechanisms face the challenge of striking a balance between competition and cooperation while preventing exploitation of the system. AI can assist by dynamically adjusting incentives based on real-time network conditions and user behavior, ensuring that participants are motivated to cooperate and use spectrum resources efficiently.

Further, privacy and security concerns are paramount, especially when AI analyzes sensitive data. To address this, advanced privacy-preserving AI techniques like federated learning and differential privacy can be implemented, safeguarding sensitive information while enabling AI-driven reputation and incentivization systems. Despite the challenges, AI integration is inevitable for continuous improvement and automation of the process and ensuring a more efficient and secure SS ecosystem in the 6G era.

9.5.4 Interoperability

Interoperability in the context of 6G networks is an inherent property, as it is an open ecosystem by design, promoting compatibility among various technology enablers. However, a significant challenge arises in establishing interoperability between different blockchain networks and traditional spectrum management systems for a secure and transparent DSS process within the 6G environment. The substantial hurdle lies in ensuring seamless interaction among different blockchain networks, legacy systems, and diverse MNOs. Overcoming this challenge necessitates the development of common protocols and data formats. To address this issue, there are various opportunities for the researchers to explore solutions such as cross-chain interoperability protocols, which serve to bridge the gap between disparate blockchains and facilitate transparent SS.

9.5.5 Churn Management in Crowdsourcing

In crowdsourcing-based transparent SS, managing participant churn while ensuring data accuracy is crucial. In this environment, participants such as MNOs may join or leave the DSS process dynamically. Handling churn while maintaining data accuracy and trust among participants is a nontrivial task. Research in this area focuses on designing robust protocols that can adapt to participant turnover while safeguarding the reliability of the shared spectrum data.

9.5.6 Design and Security Issues of Smart Contracts

The primary utility of a smart contract lies in its capacity to automate the DSS process among multiple stakeholders within the specified architecture. However, the creation of a robust and secure smart contract entails a complex endeavor. This complexity is compounded in systems where multiple blockchains have to interact for the exchange of information. In the proposed architectural setup, the task of devising and implementing secure smart contracts involves having a detailed understanding of the various functions and handling both private and public blockchain at the same time.

As the smart contracts are fundamental building block of any blockchain-based system, they must be secure and immune to any misuse. Smart contract design must incorporate best standard practices and must be thoroughly tested, verified, and reviewed. If possible, there should be a disaster recovery plan in place for any emergency. Consequently, the architecture presents a range of challenges in the secure smart contract design and deployment process.

For DSS, smart contracts are to be very fast and robust, as many real time and critical applications will run over it. Parallelly new design aspects are to be considered.

9.5.7 Distributed Interference Management with Blockchain

Interference management is one of the core components of SS. It has to be ensured that same spectrum range is not being used by more than one user in the same area. Blockchain being a distributed system and powered by smart contracts is perfectly suitable for automated and decentralized solution to the problem. However, there will be significant challenges to onboard the various diverse stakeholders on the same platform, to implement the logic in the form of smart contract and to make it run in real time.

9.6 Conclusion

In this chapter, we highlight the pivotal role of blockchain technology and smart contracts in the context of DSS and management in the next-generation wireless networks, with a specific focus on 6G networks. We delve into recent and closely related research works to identify pertinent research gaps and limitations in this field. Additionally, we provide viable blockchain and smart contract based solutions for achieving secure and trusted DSS and its management. Moreover, we discuss various challenges associated with deploying blockchain for DSS and management in B5G and 6G networks. We acknowledge that while blockchain technology shows promise in enhancing security, privacy, trust, and transparency, its integration necessitates careful consideration of trade-offs and extensive experimentation.

References

1 Panwar, N., Sharma, S., and Singh, A.K. (2016). A survey on 5G: the next generation of mobile communication. *Physical Communication* 18: 64–84.
2 Ahmad, I., Shahabuddin, S., Kumar, T. et al. (2019). Security for 5G and beyond. *IEEE Communications Surveys & Tutorials* 21 (4): 3682–3722.
3 De Alwis, C., Kalla, A., Pham, Q.-V. et al. (2021). Survey on 6G frontiers: trends, applications, requirements, technologies and future research. *IEEE Open Journal of the Communications Society* 2: 836–886.

4 Bharat 6G Vision. https://dot.gov.in/sites/default/files/Bharat%206G%20Vision%20Statement%20-%20full.pdf (accessed September 2023).

5 Saad, W., Bennis, M., and Chen, M. (2019). A vision of 6G wireless systems: applications, trends, technologies, and open research problems. *IEEE Network* 34 (3): 134–142.

6 Wu, Q., Wang, W., Li, Z. et al. (2023). SpectrumChain: a disruptive dynamic spectrum-sharing framework for 6G. *Science China Information Sciences* 66 (3): 130302.

7 Tang, F., Chen, X., Zhao, M., and Kato, N. (2022). The roadmap of communication and networking in 6G for the metaverse. *IEEE Wireless Communications* 30 (4): 72–81.

8 You, X., Wang, C.-X., Huang, J. et al. (2021). Towards 6G wireless communication networks: vision, enabling technologies, and new paradigm shifts. *Science China Information Sciences* 64: 1–74.

9 De Alwis, C., Kumar, P., Pham, Q.-V. et al. (2022). Towards 6G: key technological directions. *ICT Express* 9 (4): 525–533.

10 Patil, A., Iyer, S., Lopez, O.L.A. et al. (2022). A comprehensive survey on spectrum sharing techniques for 5G/B5G intelligent wireless networks: opportunities, challenges and future research directions. *arXiv preprint arXiv:2211.08956*.

11 Gui, G., Liu, M., Tang, F. et al. (2020). 6G: Opening new horizons for integration of comfort, security, and intelligence. *IEEE Wireless Communications* 27 (5): 126–132.

12 Kalla, A. (2023). Blockchain perspectives, mining, and types: an introductory tutorial. *IEEE Potentials* 42 (5): 23–32.

13 Kalla, A., De Alwis, C., Porambage, P. et al. (2022). A survey on the use of blockchain for future 6G: technical aspects, use cases, challenges and research directions. *Journal of Industrial Information Integration* 30: 100404.

14 Li, Z., Wang, W., Wu, Q., and Wang, X. (2022). Multi-operator dynamic spectrum sharing for wireless communications: a consortium blockchain enabled framework. *IEEE Transactions on Cognitive Communications and Networking* 9 (1): 3–15.

15 Maksymyuk, T., Gazda, J., Volosin, M. et al. (2020). Blockchain-empowered framework for decentralized network management in 6G. *IEEE Communications Magazine* 58 (9): 86–92.

16 Weiss, M.B.H., Werbach, K., Sicker, D.C., and Bastidas, C.E.C. (2019). On the application of blockchains to spectrum management. *IEEE Transactions on Cognitive Communications and Networking* 5 (2): 193–205.

17 Zhou, Y., Liu, L., Wang, L. et al. (2020). Service-aware 6G: an intelligent and open network based on the convergence of communication, computing and caching. *Digital Communications and Networks* 6 (3): 253–260.

18 Wang, S. and Sun, C. (2023). Blockchain empowered dynamic spectrum sharing: standards, state of research and road ahead. *IEEE Communications Standards Magazine* 7 (3): 72–80.

19 McLean, S. and Deane-Johns, S. (2016). Demystifying blockchain and distributed ledger technology–hype or hero? *Computer Law Review International* 17 (4): 97–102.

20 Nakamoto, S. (2008). Bitcoin: a peer-to-peer electronic cash system. *Decentralized Business Review*.

21 Saraf, C. and Sabadra, S. (2018). Blockchain platforms: a compendium. *2018 IEEE International Conference on Innovative Research and Development (ICIRD)*, 1–6. IEEE.

22 Lin, S.-Y., Zhang, L., Li, J. et al. (2022). A survey of application research based on blockchain smart contract. *Wireless Networks* 28 (2): 635–690.

23 Rouhani, S. and Deters, R. (2019). Security, performance, and applications of smart contracts: a systematic survey. *IEEE Access* 7: 50759–50779.

24 Alharby, M. and Van Moorsel, A. (2017). Blockchain-based smart contracts: a systematic mapping study. *arXiv preprint arXiv:1710.06372*.

25 Maksymyuk, T., Gazda, J., Han, L., and Jo, M. (2019). Blockchain-based intelligent network management for 5G and beyond. *2019 3rd International Conference on Advanced Information and Communications Technologies (AICT)*, 36–39. IEEE.

26 Li, Z., Wang, W., Guo, J. et al. (2022). Blockchain-empowered dynamic spectrum management for space-air-ground integrated network. *Chinese Journal of Electronics* 31 (3): 456–466.

27 Wang, S. and Sun, C. (2022). Blockchain empowered dynamic spectrum sharing: standards, state of research and road ahead. *IEEE Communications Standards Magazine* 7 (3): 72–80.

28 Gorla, P., Chamola, V., Hassija, V., and Ansari, N. (2020). Blockchain based framework for modeling and evaluating 5G spectrum sharing. *IEEE Network* 35 (2): 229–235.

29 Wang, Y., Su, Z., and Zhang, N. (2019). BSIS: Blockchain-based secure incentive scheme for energy delivery in vehicular energy network. *IEEE Transactions on Industrial Informatics* 15 (6): 3620–3631.

30 Morgado, A.J., Saghezchi, F.B., Mumtaz, S. et al. (2022). A novel machine learning-based scheme for spectrum sharing in virtualized 5G networks. *IEEE Transactions on Intelligent Transportation Systems* 23 (10): 19691–19703.

31 Matinmikko-Blue, M., Yrjölä, S., and Ahokangas, P. (2020). Spectrum management in the 6G era: the role of regulation and spectrum sharing. *2020 2nd 6G Wireless Summit (6G SUMMIT)*, 1–5.

32 Koteeshwari, R.S. and Malarkodi, B. (2022). Spectrum sensing techniques for 5G wireless networks: mini review. *Sensors International* 3: 100188.

33 Bayhan, S., Zubow, A., Gawłowicz, P., and Wolisz, A. (2019). Smart contracts for spectrum sensing as a service. *IEEE Transactions on Cognitive Communications and Networking* 5 (3): 648–660.

34 Luka, M.K., Okereke, O.U., Omizegba, E.E., and Anene, E.C. (2021). Blockchains for spectrum management in wireless networks: a survey. *arXiv preprint arXiv:2107.01005*.

10

The Synergy of Artificial Intelligence and Blockchain in 6G Spectrum Management

Ramalingam Murugan[1], *Gokul Yenduri*[2], *Pyingkodi Maran*[3], *and Thippa Reddy Gadekallu*[4]

[1] *School of Computer Science Engineering & Information Systems, Vellore Institute of Technology, Vellore Campus, Vellore, Tamil Nadu, India*
[2] *School of Computer Science and Engineering, VIT-AP University, Amaravati, Andhra Pradesh, India*
[3] *Department of Computer Applications, Kongu Engineering College, Erode, Tamil Nadu, India*
[4] *Zhongda Group, Haiyan County, Jiaxing City, Zhejiang Province, China*

10.1 Introduction

As the world prepares for the advent of 6G technology, the integration of artificial intelligence (AI) and blockchain in spectrum management emerges as a pivotal innovation. Spectrum management, the process of regulating the electromagnetic spectrum to prevent interference and optimize its use, becomes increasingly complex with the higher frequencies and broader bandwidths of 6G. Traditional methods of spectrum allocation and monitoring may fall short in addressing the dynamic and diverse requirements of 6G networks. Here, AI and blockchain technologies offer transformative solutions by enhancing the efficiency, security, and transparency of spectrum management processes.

AI, with its advanced capabilities in data analysis, machine learning, and predictive modeling, can revolutionize how spectrum resources are allocated and utilized. In the context of 6G, AI algorithms can analyze vast amounts of real-time data to predict usage patterns, detect anomalies, and optimize spectrum sharing among users. This dynamic approach allows for more efficient utilization of spectrum, reducing wastage and minimizing interference. Moreover, AI can facilitate the development of cognitive radio networks, where devices autonomously adjust their frequency bands based on real-time analysis, ensuring optimal performance and connectivity.

Intelligent Spectrum Management: Towards 6G, First Edition.
Edited by Sridhar Iyer, Anshuman Kalla, Onel Alcaraz López, and Chamitha De Alwis.

Complementing AI, blockchain technology introduces a decentralized and tamper-proof framework for spectrum management. Blockchain's distributed ledger system ensures that all transactions and allocations of spectrum resources are transparent, secure, and immutable. This is particularly critical in a 6G environment, where the sheer volume and diversity of connected devices demand robust security measures. By recording all spectrum transactions on a blockchain, stakeholders can trace and audit the allocation and usage of spectrum resources, reducing the risk of fraud and unauthorized access. Together, AI and blockchain create a synergistic effect, where AI-driven optimization and blockchain-based security and transparency converge to pave the way for an efficient, fair, and secure spectrum management system in the 6G era. Following are some of the key elements in spectrum management as depicted in Figure 10.1.

10.1.1 General Challenges of Spectrum Usage and Management

Navigating the complexities of the radio-frequency spectrum proves to be a challenging task with numerous hurdles to overcome. The limited nature of the spectrum, coupled with the constant-growing demand for wireless communication services, offers a fundamental issue of scarcity. Balancing the allocation of spectrum, reducing interference, and adapting to technological developments require cautious coordination and regulation. Moreover, spectrum is a global resource, so the international perspective of spectrum usage adds an additional layer of complexity to this critical resource management. The following are the critical challenges of spectrum management and usage are depicted in Figure 10.1.

Spectrum Scarcity [1]: The radio spectrum is a limited resource, and the demand for providing wireless communication services, such as mobile data and Wi-Fi, remains to grow. Allocating enough spectrum to meet this rising demand is a significant challenge.

Interference [1]: Spectrum bands are shared among various users and services. Interference between different devices and technologies can mess up communications. Managing these intereferences and confirming coexistence is vital.

Spectrum Allocation and Licensing [1]: Determining how to allocate spectrum to different users and services requires cautious planning and regulation. It is also difficult to balance the requirements of commercial, governmental, and public services.

Spectrum Auctions [1]: Auctioning spectrum licenses is a common process for allocating spectrum to mobile operators and ISPs. Confirming fair and efficient auctions while maximizing income for governments is a delicate balancing act.

Spectrum Policy and Regulation [1]: Developing and administering spectrum policies and regulations to prevent misuse, guarantee compliance, and encourage innovation is a continuing challenge.

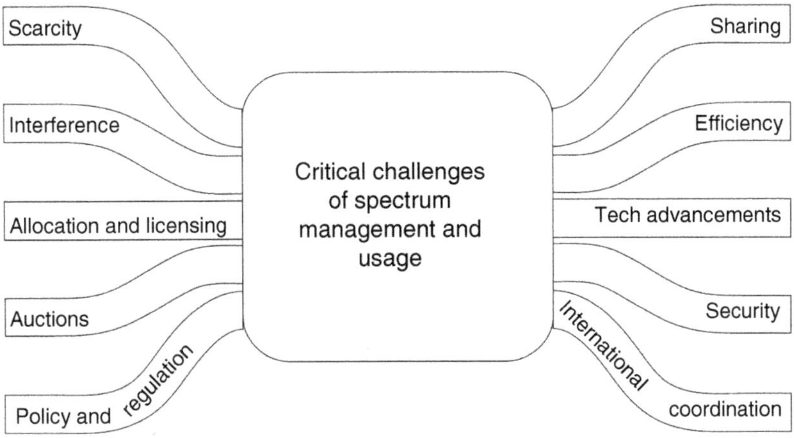

Figure 10.1 Critical challenges of 6G spectrum management and usage.

Technological Advancements [2]**:** Rapid progressions in wireless technologies, such as 5G and beyond, demand adaptive spectrum management to accommodate new evolving services and applications.

Spectrum Sharing [1]**:** Sophisticated coordination and technical solutions to avoid interference are required to effectively share spectrum among different users, like government agencies and commercial entities [3].

International Coordination [1]**:** Spectrum is a global resource, and global coordination is necessary to avoid cross-border interference and confirm compatibility between different regions.

Spectrum Efficiency [1]**:** Maximizing the efficiency of the use of spectrum through technologies like dynamic spectrum sharing, cognitive radio, and white spaces utilization is a continuing challenge.

Spectrum Security [1]**:** Restricting unauthorized access, spectrum hijacking, and cyberattacks on wireless networks is vital to maintaining the integrity of spectrum usage.

10.1.2 Importance of AI and Blockchain in Optimizing Spectrum Usage

AI plays a vital role in transforming spectrum management, the discipline focused on efficiently allocating and regulating the use of radio-frequency spectrum. Various AI techniques are used to address challenges such as interference mitigation, dynamic spectrum access, and optimal allocation.

- Machine learning (ML) algorithms, together with predictive analytics and interference detection models, analyze historical data to predict spectrum usage patterns and identify potential interferences.

- Reinforcement learning simplifies dynamic spectrum access by allowing devices to autonomously adapt their communication parameters based on real-time conditions.
- Cognitive radio systems use AI for spectrum sensing and adaptive settings, whereas game theory models optimize spectrum auctions.
- Swarm intelligence helps in collaborative spectrum sharing, and optimization algorithms discover optimal solutions for spectrum allocation.
- Natural language processing assists in policy analysis, and deep learning improves signal classification.
- Autonomous spectrum management systems make use of AI for real-time decision-making, confirming efficient and adaptive spectrum utilization.
- Fuzzy logic models handle uncertainty, offering flexibility in dynamic spectrum environments.

AI can be used to identify underutilized spectrum resources by constantly monitoring the electromagnetic spectrum. It can use its analyzing ability on its own historical data, and by using deployed machine learning algorithms to detect patterns that show underutilization of spectrum [4]. AI-enabled radios can sense and measure frequency bands in real time, while deployed predictive models can forecast future usage demands in advance. In addition, crowdsourced data and dynamic spectrum-sharing methods make use of AI to gather real-time information. This simplifies efficient spectrum access and also helps regulatory bodies and network operators optimize the utilization of this finite resource and ensure its efficient allocation for various communication services [5]. It can also be used in managing interference in spectrum by constantly monitoring the spectrum for disruptions. It can automatically detect and pinpoint possibilities of interference, and dynamically fine-tune frequency allotments and transmission parameters to alleviate disruptions in real time. Cognitive radios enabled with AI ability can intelligently shift to interference-free frequencies, while machine learning can help signal processing techniques to filter out interference [4]. AI-driven spectrum-sharing platforms can coordinate access and predict interference based on historical data. This can enhance overall interference management and help in ensuring the reliability of wireless communication services [4].

Blockchain enables dynamic spectrum access by using smart contracts for real-time allocation, maximizing utilization of available spectrum [6]. It simplifies spectrum-sharing agreements among telecom operators, IoT devices, and government bodies, leading to efficient resource use. Blockchain improves interoperability across different spectrum management systems and allows for coordinated, optimized usage through a standardized platform [7]. Its security features protect against unauthorized access and interference, ensuring legal usage. Blockchain also creates a tamper-proof record of spectrum usage, aiding in compliance monitoring and interference prevention [8].

10.2 Understanding Spectrum Management in 6G

Understanding spectrum management in 6G involves navigating the allocation and optimization of radio-frequencies to ensure seamless connectivity, minimal interference, and efficient use of the spectrum. As 6G technology aims to offer unprecedented speeds, ultralow latency, and massive device connectivity, effective spectrum management becomes crucial. This includes dynamic spectrum sharing, advanced antenna technologies, and regulatory frameworks that balance commercial, public, and private needs. Emphasizing innovation in spectrum utilization will be key to unlocking the full potential of 6G networks and supporting diverse applications, from smart cities to immersive virtual realities.

10.2.1 The Evolving Requirements and Challenges of Spectrum Management in 6G

The evolution toward 6G presents new requirements and challenges for spectrum management, driven by the demand for higher data rates, ultralow latency, and increased connectivity. With the proliferation of IoT devices, autonomous vehicles, and smart cities, spectrum resources must be managed more dynamically and efficiently [9]. Traditional static allocation methods are insufficient, necessitating advanced techniques like cognitive radio and dynamic spectrum sharing. Additionally, the integration of higher frequency bands, such as millimeter waves and terahertz frequencies, introduces new challenges in propagation, requiring sophisticated technologies to overcome issues related to range, penetration, and interference.

Security and privacy are paramount in 6G spectrum management. As networks become more complex and interconnected, ensuring the integrity and confidentiality of spectrum usage is crucial [9]. Blockchain technology can address these challenges by providing a secure, decentralized ledger for spectrum transactions, enabling real-time, tamper-proof monitoring, and compliance. Moreover, interoperability between various spectrum management systems and stakeholders, including telecom operators, IoT devices, and regulatory bodies, is essential for optimized spectrum utilization. Standardized frameworks and collaborative approaches will be critical to addressing the multifaceted requirements and challenges of spectrum management in the 6G era.

10.2.1.1 Technology Integration

Recent advancements in AI and ML have opened the door to exciting innovations like self-driving cars and voice-controlled assistants. But, 6G networks are getting more complex, so it is necessary to have smarter ways to deal with issues, manage the network, pinpoint the problems, and know how well it's performing. To handle these challenges, it is essential to integrate solutions like AI, machine learning, and software-defined networking (SDN). Machine learning and artificial intelligence are specifically important for making smart decisions to ensure the network runs smoothly. When it comes to the technical aspects, like how radio networks work, AI has the ability to make them work even better.

10.2.1.2 Security Requirements

6G networks demand strong security mechanisms to ensure spectrum usage by authorized personnel. It includes authentication and authorization techniques to confirm that only authorized users and devices can access the network. It must also use end-to-end encryption to protect data from being tampered with during transmission. This means that the data on the sender side is encrypted, and only the intended recipient can decrypt it. It must also be resilient to a variety of attacks [10]. Complex unbreakable encryption techniques to be deployed. 6G networks must also be continuously monitored for security threats. This can be done through using security tools such as network scanners and intrusion detection systems. Security must be built into the design of 6G networks from the start. This means that security must be considered at every stage of the design process, for example, in the architecture level itself, from the selection of hardware and software to the implementation of security protocols. AI will be used in 6G networks for various tasks, such as network optimization and traffic management. So, AI can also be used to launch attacks on 6G networks. So, complex security solutions that can protect AI-enabled 6G communications will be needed.

10.2.2 The Need for Efficient and Dynamic Spectrum Allocation

10.2.2.1 Escalating Demand for Bandwidth

As 6G is predicted to support a wide range of bandwidth-intensive applications such as holographic communications, autonomous vehicle communications, augmented reality, virtual reality, and ultra-high-definition content streaming, there will be an exceptional demand for spectrum resources. Effectiveness in allocation confirms that this demand can be met without congestion or performance degradation.

10.2.2.2 Diverse Use Cases

6G will tailor to a diverse set of use cases, from mission-critical applications like augmented reality (AR), virtual reality (VR) communications, autonomous vehicles communications, and remote surgery to massive IoT deployments. These applications have enormously different spectrum demands. Dynamic allocation of spectrum allows for the flexible allocation of resources based on real-time needs.

10.2.2.3 Spectrum Scarcity

Spectrum is a finite limited global resource. Many frequency bands are already allocated for various purposes. Effective utilization of available spectrum bands, including the use of previously underutilized bands or sharing spectrum among multiple services, is crucial to maximize the available resources. However, finding suitable and unallocated frequency bands for 6G poses a challenge, as the radio-frequency spectrum is a finite and shared resource. The scarcity of available spectrum for 6G can affect the deployment and effectiveness of new communication technologies, demanding careful spectrum management, allocation policies, and potentially the exploration of higher frequency bands to meet the evolving needs of 6G networks.

10.2.2.4 Minimizing Interference

Interference in 6G is influenced by a confluence of factors. The search for higher data rates, the diverse applications supported by 6G, ranging from augmented reality to massive IoT, advanced antenna technologies like massive multiple-input multiple-output (MIMO) and beamforming, dynamic spectrum sharing, network density, and connectivity, and stringent requirements for latency and reliability in applications like autonomous vehicles. As 6G prioritizes energy efficiency, interference management becomes integral for optimizing resource allocation. Security, privacy concerns, and the incorporation of AI add additional dimensions, making interference management a multifaceted challenge in the 6G. Dynamic allocation of spectrum permits the network to adapt to changing conditions. This reduces interference between different services or users. This interference management method is vibrant for maintaining network quality and reliability.

10.2.2.5 Improved Spectral Efficiency

Spectral efficiency is important in communication systems due to the limited nature of the radio-frequency spectrum. As the demand for wireless communication continues to increase with the rise in number of devices and data-intensive applications, efficient spectrum utilization becomes vital. Efficiency in spectrum utilization allows for the transmission of more data within the same bandwidth, leading to cost savings, improved energy efficiency, and the capability to serve an increasing number of users. It is mandatory to mitigate interference, enhance the quality of service, and support future technological advancements, confirming that communication systems can meet the evolving needs of an increasingly connected world. Dynamic allocation techniques of the 6G spectrum, including cognitive radio and spectrum sensing, can lead to enhanced spectral efficiency by permitting unlicensed devices to opportunistically access underutilized bands, optimizing the use of available resources.

10.3 Foundations of AI with Respect to 6G

The evolution of wireless communication technology toward 6G brings unprecedented challenges and opportunities, notably in the realm of spectrum management. As the foundation of future networks, AI plays a vital role in the efficient utilization of spectrum resources. In this section, we delve into the fundamental concepts of AI within the context of 6G spectrum management. From its role in dynamic spectrum allocation to cognitive radio systems and predictive maintenance of network infrastructure, we explore how AI-powered solutions are poised to reshape the landscape of spectrum management in the 6G era, enabling faster data rates, lower latency, and a more responsive and adaptive wireless ecosystem. Integration opportunities of AI and Blockchain in 6G spectrum management is depicted in the Figure 10.2.

10.3.1 An Overview of AI Concepts Relevant to Spectrum Management

Cognitive radio represents the first initiative using AI within the domain of spectrum management. It allows radios to independently sense and adapt to their environment by using AI algorithms. Cognitive radio can identify unused spectrum dynamically allocate frequencies, and regulate transmission parameters to avoid interference [12]. ML algorithms, such as neural networks and decision trees, can also be used to analyze spectrum data to identify unused or underutilized frequency bands [13]. This allows dynamic spectrum access and efficient spectrum sharing among multiple devices. Reinforcement learning algorithms can be used to improve spectrum management policies [14]. These algorithms study to make decisions that make the most out of network performance, diminish interference, and allocate spectrum resources effectively based on rewards and penalties. AI can also be deployed to predict future spectrum usage demands based on past usage patterns. By analyzing historical data and real-time

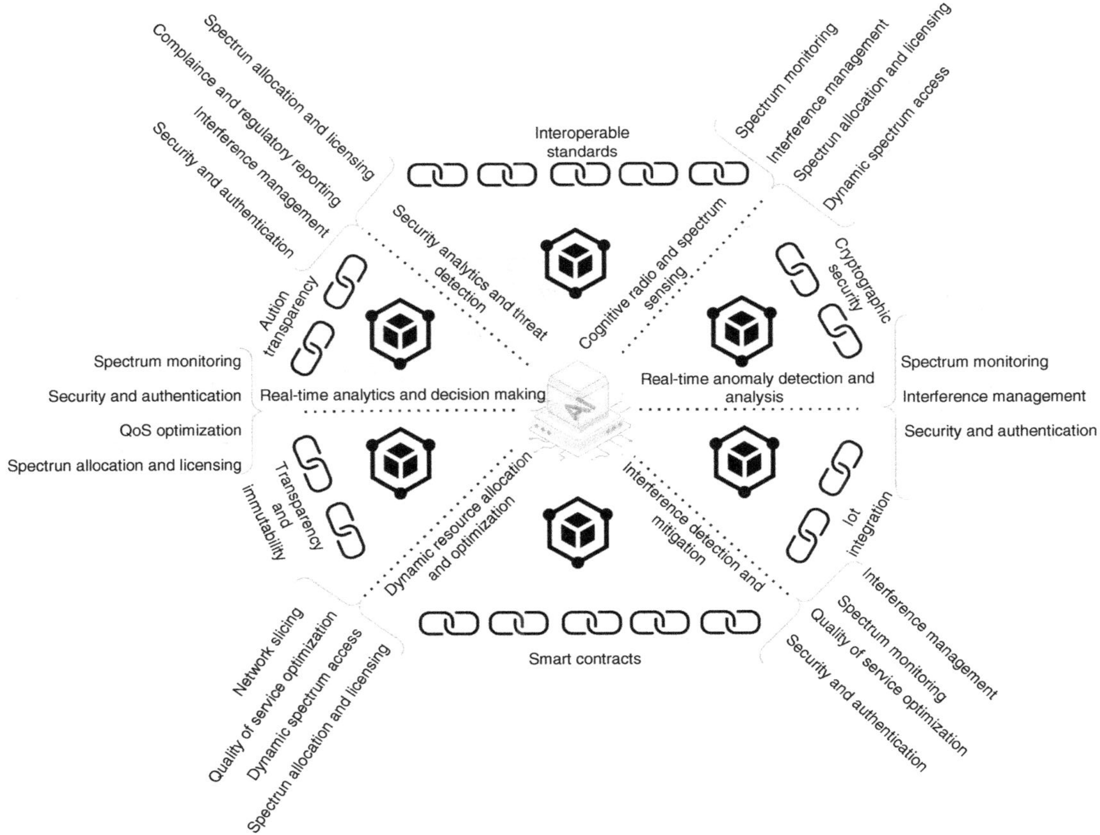

Figure 10.2 AI and blockchain for 6G spectrum management. Source: Adapted from Hu et al. [11].

information, AI models can predict and estimate when and where spectrum demand will increase, allowing preemptive spectrum allocation [15]. AI-enabled dynamic spectrum allocation algorithms can be used to continuously analyze the spectrum environment, prioritize services based on their necessities, and allocate spectrum resources accordingly. The authors in [16] have proposed AI and Blockchain integrated framework for dynamic switching of the operator and multi-operator service provision for the end users. AI can also be used to identify and alleviate interference in crowded spectrum environments. Methods such as beamforming, interference cancellation, and adaptive filtering use AI to diminish interference and improve communication quality. In situations with numerous users or devices, AI can be used to optimize resource allocation, such as time slots or frequency bands, to confirm fairness, maximize throughput, and diminish latency for all users. AI algorithms can also simplify spectrum sharing among diverse services and technologies. This ensures that they coexist pleasantly without causing harmful interference. This is crucial for integrating 6G with existing networks. A decentralized AI-based approach to spectrum management amongst multiple operators proposed in [17]. It operates based on the utilization of linear programming and a game-theoretic framework to simplify collaboration among multiple wireless service providers, enabling distributed spectrum management.

Self-organizing network (SON) technologies make use of AI to independently configure, optimize, and heal wireless networks. This includes spectrum allocation. AI-enabled SON can adapt to dynamic changing conditions and maintain consistent network performance. AI can be used to enhance the security of spectrum management by detecting unauthorized users or devices trying to access restricted frequencies. The Authors in [18] proposed an edge computing-based context-aware channel selection algorithm to perform communications with the coexistence of licensed and unlicensed spectrum. Machine learning algorithms can detect abnormalities and potential threats in the spectrum. The authors in [19] proposed ML-based conceptual framework of end-to-end learning for spectrum monitoring. The authors systematically introduced a generic methodology to easily design and implement wireless signal classifiers. AI can also be used to optimize energy consumption in devices and networks by intelligently regulating transmission power and spectrum usage based on traffic requirements and network conditions. AI can be used to support regulatory authorities in making informed decisions regarding spectrum allocation and policy enforcement. A deep reinforcement learning-based framework proposed in [20] helps achieve real-time spectrum allocation in different scenarios. Machine learning can be used to analyze large datasets to support policy-making processes in spectrum utilization and allocation [21]. On the whole, AI techniques are crucial in transforming spectrum management for 6G and future wireless networks.

They allow adaptive, efficient, and intelligent allocation of spectrum resources, confirming that the growing requirement for wireless connectivity is met while diminishing interference and enhancing network performance.

10.4 Fundamentals of Blockchain with Respect to 6G

Blockchain stands as an innovative technology that has captured considerable interest in recent times. At its core, it operates as a decentralized and distributed digital ledger, designed to securely and transparently document transactions across a network of interconnected computers.

10.4.1 Role of Blockchain in Spectrum Management

Current spectrum-sharing protocols are not automated due to the inability of authorities to monitor spectrum utilization in real time and stop unfair practices. The complexity is heightened by the large number of licensed and unlicensed bands for various bodies and wireless standards. Additionally, real-time agreements and interference management are challenging, deterring dynamic spectrum sharing. Integrating blockchain technology with the current mobile network architecture can create an automated spectrum trading framework to control MNOs' software-level spectrum usage. This framework, using smart contracts, can facilitate agreements between nearby MNOs to reduce interference issues.

Blockchain's automation, through smart contracts, can reduce the administrative workload in spectrum management by simplifying tasks like issuing licenses, handling renewals, and managing compliance reporting. Integrating IoT devices and sensors with blockchain enables real-time monitoring of spectrum usage, securely recording valuable data on the blockchain for efficient oversight by regulators and license holders [22]. Blockchain technology is crucial for optimizing wireless spectrum resources in telecommunications. However, successful integration requires collaboration among governments, regulatory bodies, telecommunications firms, and technology providers, as well as standardization initiatives and regulatory frameworks for seamless interoperability and adherence to spectrum management rules [23].

10.4.2 Enhancement of Spectrum Management Using Blockchain

Blockchain technology offers significant improvements in spectrum management by enhancing transparency, efficiency, and security in the allocation and utilization of radio-frequency (RF) spectrum resources. The primary advantage is the creation of an open, immutable ledger for spectrum allocations, recording each allocation and its terms in a transparent historical record. This transparency deters unauthorized spectrum usage and provides a reliable repository for spectrum allocation, usage, and ownership data. Unlike traditional centralized spectrum databases, blockchain-based decentralized databases are tamper-resistant, reducing the risk of manipulation or cyberattacks. Additionally, blockchain can facilitate secure and transparent spectrum trading, ensuring that transactions are accurately recorded and all relevant parties are informed. It can also support spectrum auctions by recording bids as blockchain transactions, maintaining fairness, and preventing fraud. Furthermore, blockchain's ability to authenticate and authorize devices guarantees that only legitimate devices access specific frequency bands, enhancing security against interference and unauthorized usage.

For successful implementation in 6G networks, blockchain must address high throughput and efficient storage requirements, supporting applications needing ultrahigh data rates and connection density. Solutions like sharding, directed acyclic graph (DAG), and off-chain storage are explored to increase scalability and throughput. Blockchain's immutability simplifies auditing spectrum usage, enabling real-time monitoring through IoT devices, which securely transmit data to the blockchain for effective spectrum management. This real-time data facilitates dynamic spectrum allocation based on demand, improving efficiency. Blockchain's cryptographic methods and decentralized architecture enhance security, making unauthorized alterations difficult and reducing fraud and cyberattacks. In disputes, blockchain provides a transparent, time-stamped ledger of all transactions, streamlining resolution processes. Smart contracts further enable devices to negotiate transmission parameters, minimizing interference and optimizing spectrum use.

10.5 AI-Driven Spectrum Prediction Techniques

AI-driven spectrum prediction is a technology which involves in harnessing the power of artificial intelligence and advanced machine learning methods to forecast and enhance the distribution of radio-frequency spectrum assets in support of wireless communication systems [24]. The RF spectrum, a limited and highly valuable resource, serves as the lifeblood of various wireless communication services, encompassing cellular networks, WiFi, satellite communication, and more [25]. Effective management and allocation of spectrum resources are pivotal in guaranteeing the seamless functioning of these services and addressing the growing need for wireless connectivity.

10.5.1 Data Collection

AI-powered spectrum prediction systems collect data from diverse origins, encompassing historical spectrum utilization, environmental factors, network traffic, and user requirements. This dataset serves as a critical component for training AI models. The acquisition of historical spectrum utilization data within the specified geographical area or frequency bands of interest can be sourced from a variety of outlets [26], including spectrum monitoring stations, network operators, and regulatory authorities [27]. This dataset should encompass details about the allocation of frequency bands to specific services, their usage patterns in terms of timing and location, and related information. Furthermore, it is crucial to amass environmental data that has the potential to impact spectrum propagation.

10.5.2 AI Model Training

Machine learning models, including neural networks, undergo training using the gathered dataset. These models possess the capability to discern intricate patterns and correlations within the data, enabling them to provide forecasts regarding future spectrum utilization [28]. The process of AI model training holds significant importance in the realm of AI-driven spectrum prediction. It enables machine learning models to acquire insights from historical data, facilitating the capacity to make predictions and informed decisions based on that knowledge [29]. Regular updates to the model with fresh data are imperative to ensure its adaptability to shifting conditions and evolving usage patterns. This ongoing learning process ensures the AI-driven spectrum prediction system maintains its accuracy and efficacy over time.

10.5.3 Spectrum Forecasting

Utilizing historical data and real-time conditions, the AI model possesses the capability to forecast future spectrum usage, encompassing variables like network congestion, interference, and the accessibility of spectrum across various frequency bands [30]. Spectrum forecasting serves as a pivotal element within AI-powered spectrum prediction systems. This practice leverages artificial intelligence and machine learning methodologies to predict the forthcoming utilization and availability of radio-frequency spectrum assets. Spectrum forecasting stands as a dynamic process, necessitating continuous monitoring and model updates in response to evolving circumstances. By facilitating more efficient and adaptable spectrum management, AI-driven spectrum prediction systems play a crucial role in meeting the evolving demands of wireless communication technologies and services [31].

10.5.4 Dynamic Spectrum Allocation

The concept of dynamic spectrum allocation(DSA) plays a pivotal role in the realm of AI-driven spectrum prediction and management [32]. DSA harnesses the power of artificial intelligence and machine learning methodologies to dynamically allocate and enhance the utilization of radio-frequency spectrum assets in a real-time context [33]. This approach gains particular significance in light of the escalating demand for wireless communication services, as efficient spectrum allocation holds the potential to increase capacity, diminish interference, and enhance overall network performance [34]. AI-driven spectrum prediction systems possess the capability to dynamically distribute spectrum resources to various communication services based on real-time forecasts. This allocation can be fine-tuned to ensure the efficient utilization of available spectrum while minimizing interference.

10.5.5 Cognitive Radio Systems

The integration of AI into cognitive radio systems allows them to adjust their communication parameters, including frequency, power, and modulation, based on predictions of spectrum availability and usage. This adaptation capability enhances the flexible and efficient utilization of spectrum resources [35]. Dynamic Spectrum Allocation stands as a pivotal concept within AI-driven spectrum prediction and management. DSA harnesses the power of AI and ML techniques to dynamically allocate and optimize the utilization of radio-frequency spectrum resources in real time. This approach holds particular significance in light of the increasing demand for wireless communication services, as effective spectrum allocation can result in heightened capacity, reduced interference, and overall enhanced network performance [36].

10.5.6 Spectrum Sharing

The notion of spectrum sharing holds paramount importance in the domain of AI-driven spectrum prediction and management. It revolves around the efficient allocation and cooperative use of radio-frequency spectrum resources across numerous users, services, and devices [11]. Spectrum sharing gains increasing significance due to the escalating demand for wireless communication, coupled with the fact that spectrum is becoming a finite and precious commodity. In this context, AI-driven spectrum prediction assumes a pivotal role in enabling a dynamic and efficient approach to spectrum sharing. AI can facilitate the harmonious sharing of spectrum among diverse wireless operators and services. By forecasting when and where spectrum resources will be accessible, it simplifies the coordination and equitable distribution of spectrum, mitigating the risk of harmful interference.

10.5.7 AI-Driven Spectrum Optimization Techniques

The optimization of spectrum utilization through the application of AI involves harnessing machine learning, predictive analytics, and intelligent decision-making to effectively manage and allocate radio-frequency spectrum resources. This ongoing process necessitates continuous monitoring, analysis, and adaptation, ultimately leading to enhanced spectrum management, improved resource utilization, and overall network performance enhancement within wireless communication systems [37]. Developing predictive models capable of forecasting spectrum usage patterns, identifying peak demand periods, and anticipating future interference sources serves as a cornerstone in gaining insights into spectrum utilization trends. The implementation of dynamic spectrum allocation mechanisms, driven by AI models, allows for real-time decision-making regarding the allocation of spectrum resources. These decisions take into account factors such as user demand, interference levels, and quality of service requirements. Utilizing AI to identify and counteract interference sources, whether they arise from internal network issues or external factors, is of utmost importance. The implementation of adaptive transmission parameter algorithms aimed at minimizing interference stands as a critical approach. Furthermore, deploying cognitive radio technology in tandem with AI algorithms designed for spectrum sensing empowers systems to independently recognize available spectrum bands and fine-tune transmission parameters to optimize the utilization of the spectrum. Incorporating AI-based adaptive modulation and coding schemes that adapt transmission rates based on spectrum conditions is crucial for maximizing data throughput while maintaining reliable communication. Furthermore, the utilization of AI-driven optimization algorithms ensures the efficient allocation of the appropriate amount of spectrum and power to different users and services, effectively meeting their QoS requirements [16].

10.5.8 AI for Optimization of Spectrum Utilization and Management

AI algorithms can be engaged to optimize spectrum usage in communication over wireless networks. AI algorithms influence machine learning and artificial intelligence techniques to dynamically distribute, manage, and control the radio-frequency spectrum resources competently [38].

10.5.9 Reinforcement Learning

Reinforcement learning (RL) is a prevailing method for optimizing spectrum utilization in wireless communication networks, predominantly in heterogeneous and dynamic environments where competent provision of RF spectrum resources is crucial [39]. RL permits cognitive radio systems to absorb and adapt the spectrum provision approaches based on interactions with the RF spectrum. Reinforcement learning algorithms like deep Q-networks (DQN) and Q-learning are used to assist cognitive radio systems to autonomously acquire and adapt their spectrum provision strategies based on rewards and penalties. RL agents study to make decisions that maximize long-term enactment, such as decreasing interference or increasing throughput.

10.5.10 Dynamic Programming Model

Dynamic programming is a dominant method for elevating spectrum usage in wireless communication networks, mainly in circumstances where there are complex constraints and interdependencies. Dynamic programming models can assist to find optimal spectrum provision policies that exploit utilization while following to interference, regulatory, and quality of service requirements. The objective of spectrum optimization is to distribute obtainable spectrum resources to several services, users or applications in a way that maximizes global effectiveness while satisfying constraints and diminishing interference. The dynamic programming model labeled now to provides a systematic framework for optimizing spectrum usage while considering a wide range of factors [40]. It is mainly suitable in scenarios where the spectrum environment is dynamic and efficient provision is significant for meeting the varied necessities of wireless communication services.

10.5.11 Explainable AI

Explainable AI (XAI) models are intended to offer insights into how and why certain decisions are taken by building the valuable for spectrum optimization mechanisms where transparency and accountability are significant. Transparency and interpretability in AI models are essential for optimizing the spectrum usage in particularly for the complex applications/regulated environments. Decision trees and random forest algorithms are interpretable models that can be used for forecasting spectrum utilization and provision decisions. The models are used to break down complex decisions into a sequence of simple rules that can be effortlessly understood. Decision tree nodes represent conditions based on spectrum features and leaf nodes offer the final decisions. Users can trace back through the tree to see how a decision was reached. Random Forests algorithms are ensembles of decision trees can provide improved predictive accuracy while holding interpretability. Each decision tree in the ensemble can be examined to gain understandings into the key features dynamic spectrum provisions decisions. Linear models like linear regression or logistic regression provide straightforward explanations of how various spectrum-related features influence allocation decisions. By incorporating XAI models into spectrum optimization processes, it becomes possible to ensure transparency, accountability, and regulatory compliance while still harnessing the power of advanced machine learning techniques for efficient spectrum allocation and management [41]. These models allow stakeholders to advance discernment's into how spectrum-related decisions are made and to trust the optimization process.

10.5.12 Multiagent Systems

Multiagent systems (MAS) are an efficient method for optimizing spectrum utilization in wireless communication networks, particularly in complex and dynamic environments where multiple users, devices and services coexist and participate for partial RF spectrum resources. MAS permit intelligent agents to negotiate, interact and collaborate in real time to distribute spectrum powerfully. Agents constantly manage the spectrum environment to gather real-time data on spectrum occupancy, availability, and interference levels. Agents use historical data and predictive analytics to approximation spectrum demands for different services and users. This information informs their negotiation strategies. Agents engage in negotiations to allocate spectrum resources. This can involve bidding in auctions, engaging in spectrum-sharing agreements, or making cooperative decisions to avoid interference. Agents employ interference mitigation and detection approaches to decrease interference with major users and other subordinate users [42]. The models regulate the transmission key factors like power, frequency, and modulation to reduce interference. AI-driven spectrum optimization associates the several models and approaches to enhance dynamic and real-time decisions regarding spectrum provision, interference management, and resource optimization. The best of models based on the precise goals, system requirements and environmental conditions of the communication over the wireless network.

10.6 Blockchain for Spectrum Access and Authentication

Blockchain technology can be engaged to improve spectrum access management mostly in scenarios where security, transparency, and trust among users and regulators are significant. Blockchain empowers decentralized management of spectrum resources and also decreasing the requirements for centralized spectrum authorities. Users can communicate with the blockchain to negotiate, request and attain access to the existing spectrum bands. Every spectrum provisions, sharing agreement, or transaction can be logged as a blockchain smart contract. This enables a transparent and immutable ledger of spectrum access rights, confirming trust among users and regulators.

10.6.1 Securing Spectrum Access and Authentication Using Blockchain Technologies

Blockchain technology is used for identity verification, user authentication, and certifying that only authorized user can access specific spectrum bands. This tends to decrease the risk of unauthorized or malicious use [43]. Smart contracts on the blockchain can systematize spectrum allocation and management. Devices and users are negotiating and executing the agreements based on QoS requirements, predefined rules and payment terms. Blockchain-based systems can provision dynamic spectrum access by permitting users to demand and release spectrum resources in real-time environment. This competent usage of spectrum leads to enhanced overall consumption. Users can report interference or spectrum conflicts to the blockchain network. Smart contracts can activate actions like spectrum adjustment or dispute resolution to address interference concerns promptly. Blockchain can expedite micropayments and disbursements for spectrum access. The user needs to pay for the consumption of spectrum and transactions can be logged on the blockchain, which helps for simplifying billing and revenue sharing activities. In case of disputes or conflicts on spectrum access or interference, blockchain can offered a transparent and auditable platform for dispute resolution. Smart contracts can outline procedures for resolution. All spectrum access events and authentication attempts are logged on the blockchain ledger. This audit trail delivers transparency and traceability for enabling regulators and network operators to examine security breaches or violations. Spectrum access can be managed and monitored by smart contracts that automatically perform access rights depending on predefined rules. These rules can incorporate user identity, time, location, and regulatory requirements. Unauthorized access attempts are automatically rejected by the blockchain [38]. Users and devices can have their identities verified on the blockchain through digital certificates or cryptographic signatures. The verified identities are linked to precise spectrum access permissions. Unauthorized or unverified users are incapable to attain access to spectrum resources.

10.6.2 Role of Smart Contract for Managing Spectrum Resources

Smart contracts can play a significant role in managing spectrum allocation and usage in 6G wireless communication networks. Spectrum management in 6G is an anticipated complex task and the necessity for managing dynamic resource allocation provides significant benefits from the transparency, automation, and efficiency using smart contracts [6].

Smart contracts used to automate the allocation of spectrum resources in real-time environment. When a user or device needs to access a particular frequency band, a smart contract can be activated to ensure the request, assess existing spectrum and assign the resources consequently [44]. This dynamic allocation accommodates the various and changeable demands of 6G applications. Smart contract (SC) is employed to facilitate the spectrum sharing agreements among multiple services and users. Users can negotiate then pass into smart contracts that identify how they will share the resources of the spectrum and also the conditions under which sharing occurs on any associated payments or incentives. This enables efficient use of spectrum by allowing coexistence of various services. To avoid and control interference, smart contracts can outline interference recognition and mitigation protocols. The smart contract can automatically adjust frequency bands, power levels, or access permissions to decrease the influence on neighboring services or users when the interference event is identified. Smart contracts enforce QoS necessities by identifying minimum service quality levels for various services and applications. When spectrum is assigned, the related smart contract confirms that the allocated resources meet the mentioned QoS criteria. Smart contract automatically compute fees for the usage and starts micropayments between service providers and users to certify timely settlement. This streamlines the financial facets of spectrum resource management. It controls and manages an immutable record of all spectrum allocation and usage transaction details on the blockchain. This ledger delivers transparency and serves as an auditable source of truth for operators, regulators, and users. Authenticate the identities of devices and users for ensuring that only the authorized entities can access the spectrum resources from the wireless network communication. Identity management and authentication protocols can be implanted within smart contracts for the effective and efficient management of available spectrum resource [45]. Smart contracts can employ optimization techniques to assign available spectrum resources efficiently based on interference levels, demand, and QoS requirements. Smart contracts adapt to fluctuating network conditions and prioritize high-priority applications or users. In cases of disputes over spectrum access or interference issues, smart contracts can describe procedures for determination. Dispute resolution techniques are preprogrammed by ensuring a rational and transparent process. Smart contracts operate in real-time for making prompt decisions and alterations based on the existing state of the network. This adaptability is important for managing and controlling the dynamic nature of 6G spectrum [46]. Smart contracts offer an automated, robust and transparent techniques for spectrum management in

6G networks and Smart contracts empowering competent spectrum allocation, interference management, compliance with regulatory, and QoS requirements.

10.7 Synergy of AI and Blockchain in 6G

The possible convergence of blockchain technology (BT) and AI within the framework of 6G technology, based on broad patterns as of my previous update. Blockchain technology has the potential to enhance communication reliability, lower latency, and improve security. Blockchain's built-in security features, like consensus procedures and cryptographic hashing, can be used to improve 6G network security. This is especially crucial because 6G is anticipated to manage enormous volumes of data, including private data from numerous linked devices. Smart contracts, which are self-executing agreements written directly into code, can be used in 6G networks to automate specific tasks [47]. They can enable network slicing for particular applications, dynamic spectrum allocation, or even automatic billing. Blockchain's immutability and transparency can be used to guarantee data integrity in 6G networks. For applications like industrial automation, driverless cars, and healthcare that demand high data accuracy, this is essential. Tokenization—the representation of real-world assets on a blockchain—is made possible by blockchain. This might result in new business models for 6G, where users can exchange bandwidth and easily make micropayments between devices, making the network more effective and dynamic. AI algorithms can be used to examine the enormous volume of data produced by 6G networks. Blockchain can guarantee the accuracy of the data that these AI algorithms rely on, giving a safe basis for enhancing resource allocation, predictive maintenance, and network performance optimization [37]. In 6G networks, blockchain technology can be utilized for safe, decentralized identity management. By fostering trust between devices and other network entities, this can lower the possibility of identity-related attacks. It is noteworthy that the effective amalgamation of AI and blockchain technology in 6G networks necessitates cooperation between industry participants, standardization endeavors, and continuous investigation to tackle obstacles like scalability, interoperability, and energy efficiency.

10.7.1 A Deep Dive of AI Integration with Blockchain for 6G

Enhancing the efficiency, security, and flexibility of spectrum distribution and management can be achieved through the integration of AI with blockchain technology in the context of 6G spectrum management. In order to successfully integrate blockchain technology and AI into 6G spectrum management, telecoms companies, government agencies, and technology suppliers must work together. Furthermore, essential to guaranteeing compatibility and interoperability amongst various systems would be standardization activities.

In order to dynamically assign spectrum resources, AI systems can assess real-time data, such as network traffic, user demand, and environmental factors. This may result in utilizing available frequencies in an optimal and effective manner. With blockchain technology, trust and security can be maintained by transparently recording and managing these dynamic allocations [16]. The spectrum auction procedure can be automated using smart contracts built on blockchain. In order to help determine the best starting bids and reserve prices, AI algorithms can be linked to assess past data and demand patterns. Smart contracts on the blockchain enable the complete auction process to be carried out automatically, from bidding to payment settlement. AI has the capacity to forecast 6G spectrum problems, including interference and congestion. AI algorithms can proactively optimize spectrum consumption by continuously monitoring and analyzing data. The blockchain enables the recording of decisions and results via AI-driven optimization algorithms, offering a transparent and auditable history of spectrum management [48].

The decentralized nature of blockchain technology can be utilized to allocate spectrum management responsibilities across several network nodes. These nodes' AI techniques are able to jointly decide how to allocate spectrum depending on local observations. By guaranteeing that these choices are documented in an unchangeable ledger, blockchain fosters transparency, and inhibits unlawful modifications. Spectrum resources can be tokenized using blockchain, which turns them into digital assets that are tracked on the network. The trading and leasing of these tokenized spectrum resources, depending on supply and demand can subsequently be facilitated by AI algorithms. This is a more adaptable and market-oriented method of managing spectrum.

Blockchain networks can be equipped with AI-powered security measures to identify and stop malicious activity or unwanted access pertaining to the 6G spectrum. AI analytics can track network performance over time, pinpointing problem areas and modifying spectrum distribution accordingly. Blockchain technology has the ability to log and date these modifications, producing an open, verifiable history of network changes.

10.7.2 Amalgamation of Blockchain with AI for 6G Spectrum

Transparency, security, and efficiency of spectrum management can be improved by integrating blockchain technology with AI for 6G spectrum management.

A more safe, open, and cooperative method of managing spectrum resources is encouraged by the incorporation of blockchain technology into AI for 6G spectrum management. To ensure the practical implementation of such integrated systems in actual 6G networks, it is imperative to tackle issues like scalability and interoperability [49].

The decentralized ledger of blockchain technology guarantees that information pertaining to spectrum management decisions is unchangeable once it is entered. This feature increases confidence in the integrity and accuracy of the data that AI algorithms use. Blockchain-based smart contracts have the ability to automate and enforce spectrum transaction agreements. Spectrum rights may be leased, traded, or auctioned off. The blockchain ensures that these transactions are transparent. Blockchain technology enables AI algorithms to record their results for spectrum allocation, optimization, and prediction. This produces a decentralized, transparent record of the decision-making procedures [50].

Secure identity management using blockchain ensures that only authorized parties can access spectrum resources. This is especially crucial for 6G networks since a wide range of devices and services need to use the spectrum [48]. On the blockchain, spectrum resources can be tokenized to represent them as digital assets. Based on current demand and usage trends, AI algorithms can then optimize the distribution and trading of these tokenized spectrum resources. Every decision and transaction made on the blockchain can be tracked down and verified. This helps ensure that spectrum management procedures follow set protocols and comply with regulations. Blockchain enables decentralized cooperation between AI nodes in charge of spectrum management. Information can be shared amongst various AI algorithms on the network, facilitating more cooperative and knowledgeable spectrum management [51].

10.8 Conclusion

In conclusion, the amalgamation of AI and blockchain technologies into 6G spectrum management denotes a paradigm shift in the way we optimize and secure wireless communication resources. AI's real-time analysis empowers dynamic spectrum allocation and interference detection, allowing 6G networks to respond quickly to fluctuating demands. Blockchain, with its decentralized ledger and smart contract abilities, confirms trustless, transparent, and secure spectrum allocation. Together, these technologies take to a future where spectrum resources are used effectively, sharing is seamless and secured, and interference is minimized. This dynamic combination not only promises to improve connectivity and data transfer speeds but also strengthens the integrity and security of 6G networks, setting them as the backbone of the digital age.

References

1 Attar, A., Ghorashi, S.A., Sooriyabandara, M., and Aghvami, A.H. (2008). Challenges of real-time secondary usage of spectrum. *Computer Networks* 52 (4): 816–830.

2 Ahmad, I., Shahabuddin, S., Kumar, T. et al. (2019). Security for 5G and beyond. *IEEE Communications Surveys & Tutorials* 21 (4): 3682–3722.

3 Giordani, M., Polese, M., Mezzavilla, M. et al. (2020). Toward 6G networks: use cases and technologies. *IEEE Communications Magazine* 58 (3): 55–61.

4 Li, Z., Ding, Z., Shi, J. et al. (2020). Guest editorial: artificial intelligence (AI)-driven spectrum management. *China Communications* 17 (2): iii–v.

5 Letaief, K.B., Chen, W., Shi, Y. et al. (2019). The roadmap to 6G: AI empowered wireless networks. *IEEE Communications Magazine* 57 (8): 84–90.

6 Weiss, M.B., Werbach, K., Sicker, D.C., and Bastidas, C.E.C. (2019). On the application of blockchains to spectrum management. *IEEE Transactions on Cognitive Communications and Networking* 5 (2): 193–205.

7 Liang, Y.-C. and Liang, Y.-C. (2020). Blockchain for dynamic spectrum management. In: Emre Celebi, Jingdong Chen, E. S. Gopi, Amy Neustein, Antonio Liotta, Mario Di Mauro *Dynamic Spectrum Management: From Cognitive Radio to Blockchain and Artificial Intelligence*. Springer, 121–146.

8 Yang, L., Zhang, Z., Zhao, B.Y. et al. (2012). Enforcing dynamic spectrum access with spectrum permits. *Proceedings of the 13th ACM International Symposium on Mobile Ad Hoc Networking and Computing*, 195–204.

9 Viswanathan, H. and Mogensen, P.E. (2020). Communications in the 6G era. *IEEE Access* 8: 57063–57074.

10 Ramalingam, M., Selvi, G.C., Victor, N. et al. (2023). A comprehensive analysis of blockchain applications for securing computer vision systems. *IEEE Access* 11: 107309–107330.

11 Hu, S., Liang, Y.-C., Xiong, Z., and Niyato, D. (2021). Blockchain and artificial intelligence for dynamic resource sharing in 6G and beyond. *IEEE Wireless Communications* 28 (4): 145–151.

12 Wang, B. and Liu, K.R. (2010). Advances in cognitive radio networks: a survey. *IEEE Journal of Selected Topics in Signal Processing* 5 (1): 5–23.

13 Almazrouei, E., Gianini, G., Almoosa, N., and Damiani, E. (2020). What can machine learning do for radio spectrum management?. *Proceedings of the 16th ACM Symposium on QoS and Security for Wireless and Mobile Networks*, 15–21.

14 Di Felice, M., Bedogni, L., and Bononi, L. (2019). Reinforcement learning-based spectrum management for cognitive radio networks: a literature review and case study. In: Zhang, W. (eds). *Handbook of Cognitive Radio*, vol. 3, 1849–1886. Singapore: Springer.

15 Srivastava, G., Jhaveri, R.H., Bhattacharya, S. et al. (2022). XAI for cybersecurity: state of the art, challenges, open issues and future directions. *arXiv preprint arXiv:2206.03585.*

16 Maksymyuk, T., Gazda, J., Liyanage, M. et al. (2021). AI-enabled blockchain framework for dynamic spectrum management in multi-operator 6G networks. In: Klymash, M., Beshley, M., Luntovskyy, A. (eds). *Future Intent-Based Networking: On the QoS Robust and Energy Efficient Heterogeneous Software Defined Networks*, 322–338. Springer.

17 Hlophe, M.C. and Maharaj, B.T. (2021). AI meets CRNs: a prospective review on the application of deep architectures in spectrum management. *IEEE Access* 9: 113954–113996.

18 Liao, H., Chen, X., Zhou, Z. et al. (2020). Licensed and unlicensed spectrum management for cognitive M2M: a context-aware learning approach. *IEEE Transactions on Cognitive Communications and Networking* 6 (3): 915–925.

19 Kulin, M., Kazaz, T., Moerman, I., and De Poorter, E. (2018). End-to-end learning from spectrum data: a deep learning approach for wireless signal identification in spectrum monitoring applications. *IEEE Access* 6: 18484–18501.

20 Lei, W., Ye, Y., and Xiao, M. (2020). Deep reinforcement learning-based spectrum allocation in integrated access and backhaul networks. *IEEE Transactions on Cognitive Communications and Networking* 6 (3): 970–979.

21 Srivastava, G., K, D.R., Yenduri, G. et al. (2023). Federated learning enabled edge computing security for internet of medical things: concepts, challenges and open issues. In: Srivastava, G., Ghosh, U., Lin, J.CW. (eds). *Security and Risk Analysis for Intelligent Edge Computing*, 67–89. Springer.

22 Brilliantova, V. and Thurner, T.W. (2019). Blockchain and the future of energy. *Technology in Society* 57: 38–45.

23 Qiu, J., Grace, D., Ding, G. et al. (2019). Blockchain-based secure spectrum trading for unmanned-aerial-vehicle-assisted cellular networks: an operator's perspective. *IEEE Internet of Things Journal* 7 (1): 451–466.

24 Pandya, S., Srivastava, G., Jhaveri, R. et al. (2023). Federated learning for smart cities: a comprehensive survey. *Sustainable Energy Technologies and Assessments* 55: 102987.

25 Song, H., Bai, J., Yi, Y. et al. (2020). Artificial intelligence enabled Internet of Things: network architecture and spectrum access. *IEEE Computational Intelligence Magazine* 15 (1): 44–51.

26 Yenduri, G., Srivastava, G., Maddikunta, P.K.R. et al. (2023). Generative pre-trained transformer: a comprehensive review on enabling technologies, potential applications, emerging challenges, and future directions. *arXiv preprint arXiv:2305.10435.*

27 Zhang, S. and Zhu, D. (2020). Towards artificial intelligence enabled 6G: state of the art, challenges, and opportunities. *Computer Networks* 183: 107556.

28 Wahab, O.A., Mourad, A., Otrok, H., and Taleb, T. (2021). Federated machine learning: survey, multi-level classification, desirable criteria and future

directions in communication and networking systems. *IEEE Communications Surveys & Tutorials* 23 (2): 1342–1397.

29 Yang, H., Alphones, A., Xiong, Z. et al. (2020). Artificial-intelligence-enabled intelligent 6G networks. *IEEE Network* 34 (6): 272–280.

30 AbdulRahman, S., Tout, H., Ould-Slimane, H. et al. (2020). A survey on federated learning: the journey from centralized to distributed on-site learning and beyond. *IEEE Internet of Things Journal* 8 (7): 5476–5497.

31 Berardinelli, G., Mogensen, P., and Adeogun, R.O. (2020). 6G subnetworks for life-critical communication. *2020 2nd 6G Wireless Summit (6G SUMMIT)*, 1–5. IEEE.

32 Rahman, S.A., Tout, H., Talhi, C., and Mourad, A. (2020). Internet of things intrusion detection: centralized, on-device, or federated learning?. *IEEE Network* 34 (6): 310–317.

33 Hammoud, A., Otrok, H., Mourad, A., and Dziong, Z. (2022). On demand fog federations for horizontal federated learning in IoV. *IEEE Transactions on Network and Service Management* 19 (3): 3062–3075.

34 Maksymyuk, T., Gazda, J., Volosin, M. et al. (2020). Blockchain-empowered framework for decentralized network management in 6G. *IEEE Communications Magazine* 58 (9): 86–92.

35 Tran, H.-V., Kaddoum, G., Elgala, H. et al. (2020). Lightwave power transfer for federated learning-based wireless networks. *IEEE Communications Letters* 24 (7): 1472–1476.

36 Yang, P., Kong, L., and Chen, G. (2021). Spectrum sharing for 5G/6G URLLC: research frontiers and standards. *IEEE Communications Standards Magazine* 5 (2): 120–125.

37 Zuo, Y., Guo, J., Gao, N. et al. (2023). A survey of blockchain and artificial intelligence for 6G wireless communications. *IEEE Communications Surveys & Tutorials*. 25 (4): 2494–2528.

38 Matinmikko-Blue, M., Yrjölä, S., and Ahokangas, P. (2020). Spectrum management in the 6G era: the role of regulation and spectrum sharing. *2020 2nd 6G Wireless Summit (6G SUMMIT)*, 1–5, IEEE.

39 Naparstek, O. and Cohen, K. (2018). Deep multi-user reinforcement learning for distributed dynamic spectrum access. *IEEE Transactions on Wireless Communications* 18 (1): 310–323.

40 Mafakheri, F., Breton, M., and Ghoniem, A. (2011). Supplier selection-order allocation: a two-stage multiple criteria dynamic programming approach. *International Journal of Production Economics* 132 (1): 52–57.

41 Brik, B., Chergui, H., Zanzi, L. et al. (2023). A survey on explainable ai for 6G O-RAN: architecture, use cases, challenges and research directions. *arXiv preprint arXiv:2307.00319*.

42 Jaishanthi, B., Ganesh, E., and Sheela, D. (2019). Priority-based reserved spectrum allocation by multi-agent through reinforcement learning in cognitive radio network. *Automatika* 60 (5): 564–569.

43 Kumari, A., Gupta, R., and Tanwar, S. (2021). Amalgamation of blockchain and IoT for smart cities underlying 6G communication: a comprehensive review. *Computer Communications* 172: 102–118.

44 Li, X., Jiang, P., Chen, T. et al. (2020). A survey on the security of blockchain systems. *Future Generation Computer Systems* 107: 841–853.

45 Xu, H., Klaine, P.V., Onireti, O. et al. (2020). Blockchain-enabled resource management and sharing for 6G communications. *Digital Communications and Networks* 6 (3): 261–269.

46 Ji, C. and Zhu, Q. (2021). Smart contract-based secure cooperative spectrum sensing algorithm. *International Journal of Distributed Sensor Networks* 17 (12). https://doi.org/10.1177/15501477211058673.

47 Atlam, H.F., Azad, M.A., Altamimi, M., and Fadhel, N. (2022). Role of blockchain and AI in security and privacy of 6G. In: Dutta Borah, M., Singh, P., Deka, G.C. (eds). *AI and Blockchain Technology in 6G Wireless Network*, 93–115. Springer.

48 Li, W., Su, Z., Li, R. et al. (2020). Blockchain-based data security for artificial intelligence applications in 6G networks. *IEEE Network* 34 (6): 31–37.

49 Kalla, A., De Alwis, C., Porambage, P. et al. (2022). A survey on the use of blockchain for future 6G: technical aspects, use cases, challenges and research directions. *Journal of Industrial Information Integration* 30: 100404.

50 Patel, F., Bhattacharya, P., Tanwar, S. et al. (2021). Block6Tel: Blockchain-based spectrum allocation scheme in 6G-envisioned communications. *2021 International Wireless Communications and Mobile Computing (IWCMC)*, 1823–1828. IEEE.

51 Pathak, V., Pandya, R.J., Bhatia, V., and Lopez, O.A. (2023). Qualitative survey on artificial intelligence integrated blockchain approach for 6G and beyond. *IEEE Access* 11: 105935–105981.

11

Conclusions

In view of the ongoing extensive research on BG5/6G networks and corresponding spectrum management solutions, this book provides a timely and comprehensive understanding of the research and development aspects of spectrum management for these advanced systems. Initially, the evolution of mobile networks was overviewed, followed by the spectrum management technologies, which included the principles, technical aspects, applications, and challenges. Further, the book discussed the evolution of spectrum management toward 6G and focused on the infrastructure required for DSM. Various technologies, such as deep learning, reinforcement learning, and blockchain, have been identified and discussed. The book also discussed aspects related to the convergence of AI and blockchain for spectrum management. Specifically, the mutual benefits, challenges, and prospective research on AI and blockchain for spectrum management toward 6G were presented. Lastly, the book discussed aspects related to physical and virtual infrastructure for spectrum management.

The book detailed that there are four options to access the spectrum, namely using local licenses from the regulator, spectrum acquired from the incumbent spectrum users, new spectrum brokerage model, and unlicensed access. These options differ in characteristics in terms of spectrum management models, spectrum sharing, and spectrum valuation and pricing. All these may benefit from AI for the efficient operation of B5G/6G networks. Further, 6G will significantly benefit from DSM, and as more access systems are available, 6G will be more flexible in spectrum and network adaptability. Note that realizing the 6G goals of "connect the unconnected" will require automated installation, simple management, and maintenance wherein, automated spectrum management through DSM will be a key solution. It is established that there is a pressing need for a *world model* encoding the background knowledge and experiences with the capability of planning and reasoning for which, generative AI and LLM will play a key role. The use of blockchain technology, along with smart contracts, can ensure transparent spectrum utilization, dynamic and trustless agreement for spectrum

Intelligent Spectrum Management: Towards 6G, First Edition.
Edited by Sridhar Iyer, Anshuman Kalla, Onel Alcaraz López, and Chamitha De Alwis.

sharing, tokenization of spectrum, secure auditing, and decentralized and secure overall spectrum management in 6G. Moreover, the promising convergence of blockchain and AI for spectrum management in 6G is also explored in this book. The proper exploitation of these technologies will enable a future where, spectrum resources are used effectively, sharing is seamless and secured, and interference is minimized. This dynamic combination not only promises to improve connectivity and data transfer speeds but also strengthens the integrity and security of 6G networks, setting them as the backbone of the digital age.

Significant research efforts are needed to improve both, the technical aspects and use cases of the 6G ecosystem. Specifically, the four spectrum access options detailed in the book require further research in terms of different techniques for managing the resulting interference scenarios, potential spectrum bands for local 6G networks, suitable spectrum awarding models, and spectrum pricing and valuation methods, which are in the infancy for local networks. The future spectrum access models will probably be a mix of those discussed in this book. The pathway to an AI-native wireless communication system with DSM is still challenging since there is a large gap between the theoretical performance and the practical uses with severe requirements of reliability, energy efficiency, and implementation cost. Therefore, significant advancements in research and development of both, AI and telecom systems are needed to realize a fully autonomous dynamic spectrum management system. In regard to RIS-SWIPT-enabled 6G wireless communication systems, much research is required in terms of optimal RIS placement, channel estimation, wireless power transfer efficiency optimization, and RIS selection. Further, while blockchain technology shows promise in enhancing security, privacy, trust, and transparency, its integration necessitates careful consideration of trade-offs and extensive experimentation. Lastly, note that spectrum management, characterized by the principles of spectrum virtualization and network slicing, is on the verge of a revolutionary period in telecommunications. Hence, effective spectrum management, supported by virtualization and network slicing, will be crucial in guaranteeing connectivity, stability, and innovation.

In essence, this book provided a comprehensive reference on spectrum management techniques for B5G/6G intelligent wireless networks.

Index

Page numbers with "*f*" refer to figures and those with "*t*" refer to tables.

Intelligent Spectrum Management: Towards 6G, First Edition.
Edited by Sridhar Iyer, Anshuman Kalla, Onel Alcaraz López, and Chamitha De Alwis.

f

O

z

www.ingramcontent.com/pod-product-compliance
Lightning Source LLC
Chambersburg PA
CBHW060604230126
37971CB00003B/108
* 9 7 8 1 3 9 4 2 0 1 2 0 4 *